AF389398

Cognitive Robotics & Control

Cognitive Robotics & Control

Special Issue Editor

Cecilio Angulo

MDPI • Basel • Beijing • Wuhan • Barcelona • Belgrade • Manchester • Tokyo • Cluj • Tianjin

Special Issue Editor
Cecilio Angulo
Universitat Politècnica de Catalunya—BarcelonaTech
Spain

Editorial Office
MDPI
St. Alban-Anlage 66
4052 Basel, Switzerland

This is a reprint of articles from the Special Issue published online in the open access journal *Electronics* (ISSN 2079-9292) (available at: https://www.mdpi.com/journal/electronics/special_issues/cong_robotics).

For citation purposes, cite each article independently as indicated on the article page online and as indicated below:

LastName, A.A.; LastName, B.B.; LastName, C.C. Article Title. *Journal Name* **Year**, *Article Number*, Page Range.

ISBN 978-3-03936-246-2 (Hbk)
ISBN 978-3-03936-247-9 (PDF)

Contents

About the Special Issue Editor

Cecilio Angulo BSc/MSc in Mathematics from the University of Barcelona, Spain, 1993, and PhD in Sciences from the Universitat Politècnica de Catalunya (UPC) in 2001. He joined the UPC in 1999 and is full professor of Artificial Intelligence and Robotics. Head Director of the Research Centre on Intelligent Data Science and Artificial Intelligence (IDEAI-UPC). He has worked on theoretical aspects on cognition, reinforcement learning, engineering control and robotics and applications on control systems, cognitive social robots and assistive technologies. He has authored books in machine learning and robots, and published more than 275 papers in international and national journals and conferences. He has led and participated in 35 R&D competitive projects, 12 of them funded by the European Commission.

 *electronics*

Editorial

Cognitive Robotics and Control

Cecilio Angulo

IDEAI-UPC, Universitat Politècnica de Catalunya, 08034 Barcelona, Spain; cecilio.angulo@upc.edu;
Tel.: +34-9341-34021

Received: 29 April 2020; Accepted: 1 May 2020; Published: 6 May 2020

1. Introduction

Robotics and control are both research and application domains that have been frequently engineered with the use of interdisciplinary approaches like cybernetics [1]. Cognition is a particular concept of this approach, abstracted from the context of living organisms to that of artificial devices, about knowledge acquisition and understanding through thought, experience, and the senses [2]. Cognitive robotics and control refer to knowledge processing as much as knowledge generation from problem understanding, leading to special forms of architectures enabling systems to behave in an autonomous way [3–5].

The main aim of this special issue is to highlight emerging applications and address recent breakthroughs in the domain of cognitive robotics and control and related areas. Procedures, algorithms, architectures and implementations for reasoning, problem solving or decision making in the domain of robotics and control are elements under consideration.

2. The Present Issue

This special issue consists of eight papers covering important topics in the field of cognitive robotics and control, including robotic platforms in interactive scenarios such as operating rooms, trajectories learning and optimisation from nature-inspired and computational cognition approaches, and hardware developments for motor control. The contents of these papers are introduced here.

Robotic platforms are taking their place in the operating room, providing stability and accuracy during surgery. Most of these platforms are tele-operated, as in Reference [6], where the learning from demonstration (LfD) approach is extended for object tele-manipulation. The method is experimentally verified in a tele-operated task using a lightweight robot remotely controlled with a haptic device. In the same domain, research is also being carried out to design collaborative platforms, reducing surgeon workload. The automation of auxiliary tasks would benefit both surgeons and patients by facilitating the surgery and reducing the operation time. A novel autonomous camera guidance approach for laparoscopic surgery is proposed in Reference [7], using LfD as well as being validated using an experimental surgical robotic platform. Moving forward, an important step towards a more natural and user-friendly manner of physical human-robot interaction in scenarios where humans and robots collaborate in the accomplishment of a task is presented in Reference [8]. A robotic system is introduced that is able to identify different humans' intentions and to adapt its behaviour consequently, only employing force data.

Nature-inspired solutions, like particle swarm optimisation (PSO) and artificial bee colony (ABC), are employed in Reference [9] as meta-heuristic optimisation techniques to tune a proportional-integral-derivative (PID) controller for an upper limb rehabilitation robotic arm exoskeleton RAX-1. In a different way, computational solutions are based on ontologies and knowledge representation. Aiming to represent the knowledge in robot task planning, the Robot Task Planning Ontology (RTPO) is first designed and implemented in Reference [10], so that robots can understand and know how to carry out task planning to reach the goal state. Experimental results demonstrate good performance in scalability and responsiveness. In Reference [11], the authors focus on the challenging problem of trajectory optimisation for an automatic spraying robot. Using the Bézier surface approach, an automatic solution is provided in the form of an initial trajectory, establishing the appropriate spraying model, planing the appropriate space path, and finally planing the trajectory optimisation along the specified painting path.

Field-Programmable Gate Arrays (FPGA) are considered in Reference [12] as a balanced approach to develop computing in technological low-cost applications empowered with the flexibility of software and the high-speed operation of hardware. A robotics application to control an inverted pendulum robot is designed, built, and programmed using open FPGA tools. In Reference [13], the rate of change in acceleration value is used to develop an S-curve velocity profile for motion control, which presents smoother movements to avoid high stress in the motor than in the trapezoidal velocity profile. The new methodology is developed applying an open source architecture in a hybrid electronic platform compounded by a system on a chip (SoC) Raspberry Pi 3 and a FPGA.

3. Future Research

Socio-technical system (STSs) are those considering requirements spanning hardware, software, personal, and community aspects [14–16]. Some examples were shown in our special issue, from operating rooms and exoskeletons to industrial robots for automatic spraying and motion control using edge computing. In this kind of environment, assistance to skilled users (workers, health professionals, impaired people) becomes crucial. Beyond ergonomic or safety issues, new qualification and technical competences with regard to users are needed. Hence, a hot topic of research activity to look at in the near future for robotics and control is to consider the cognitive or social dimension, by the development of computational agents, robots or electronic devices in the edge designed for increasing efficiency and effectiveness in the environment and global organisation [17].

Funding: This research has been co-financed by the European Regional Development Fund of the European Union in the framework of the ERDF Operational Program of Catalonia 2014-2020 with a grant of 1,331,903.77 € grant number 001-P-001643.

Acknowledgments: I would like to thank all researchers who submitted articles to this special issue for their excellent contributions. I also sincerely appreciate all valuable comments and suggestions from the reviewers, which helped to improve the quality of the article. I would like to acknowledge the editorial board of Electronics, who invited me to guest edit this special issue. I'm also grateful to the Electronics Editorial Office staff who worked thoroughly to maintain the rigorous peer-review schedule and timely publication.

Conflicts of Interest: The authors declare no conflict of interest. The funder had no role in the design of the issue, in selection of contributions, in the writing of the editorial, or in the decision to publish this issue.

References

1. Wiener, N. *Cybernetics, Second Edition: Or the Control and Communication in the Animal and the Machine*; The MIT Press: Cambridge, MA, USA, 1965.
2. Lazzeri, N.; Mazzei, D.; Cominelli, L.; Cisternino, A.; De Rossi, D. Designing the Mind of a Social Robot. *Appl. Sci.* **2018**, *8*, 302. [CrossRef]
3. Puigbo, J.Y.; Pumarola, A.; Angulo, C.; Tellez, R. Using a cognitive architecture for general purpose service robot control. *Connect. Sci.* **2015**, *27*, 105–117. [CrossRef]
4. Pfeiffer, S.; Angulo, C. Gesture learning and execution in a humanoid robot via dynamic movement primitives. *Pattern Recognit. Lett.* **2015**, *67*, 100–107. [CrossRef]
5. Acevedo-Valle, J.M.; Angulo, C.; Moulin-Frier, C. Autonomous Discovery of Motor Constraints in an Intrinsically Motivated Vocal Learner. *IEEE Trans. Cognit. Dev. Syst.* **2018**, *10*, 314–325. [CrossRef]
6. Pérez-del Pulgar, C.J.; Smisek, J.; Rivas-Blanco, I.; Schiele, A.; Muñoz, V.F. Using Gaussian Mixture Models for Gesture Recognition During Haptically Guided Telemanipulation. *Electronics* **2019**, *8*, 772. [CrossRef]
7. Rivas-Blanco, I.; Perez-del Pulgar, C.J.; López-Casado, C.; Bauzano, E.; Muñoz, V.F. Transferring Know-How for an Autonomous Camera Robotic Assistant. *Electronics* **2019**, *8*, 224. [CrossRef]

8. Olivares-Alarcos, A.; Foix, S.; Alenyà, G. On Inferring Intentions in Shared Tasks for Industrial Collaborative Robots. *Electronics* **2019**, *8*, 1306. [CrossRef]

9. Joyo, M.K.; Raza, Y.; Ahmed, S.F.; Billah, M.M.; Kadir, K.; Naidu, K.; Ali, A.; Mohd Yusof, Z. Optimized Proportional-Integral-Derivative Controller for Upper Limb Rehabilitation Robot. *Electronics* **2019**, *8*, 826. [CrossRef]

10. Sun, X.; Zhang, Y.; Chen, J. RTPO: A Domain Knowledge Base for Robot Task Planning. *Electronics* **2019**, *8*, 1105. [CrossRef]

11. Chen, W.; Liu, J.; Tang, Y.; Ge, H. Automatic Spray Trajectory Optimization on Bézier Surface. *Electronics* **2019**, *8*, 168. [CrossRef]

12. Ordóñez Cerezo, J.; Castillo Morales, E.; Cañas Plaza, J.M. Control System in Open-Source FPGA for a Self-Balancing Robot. *Electronics* **2019**, *8*, 198. [CrossRef]

13. García-Martínez, J.R.; Rodríguez-Reséndiz, J.; Cruz-Miguel, E.E. A New Seven-Segment Profile Algorithm for an Open Source Architecture in a Hybrid Electronic Platform. *Electronics* **2019**, *8*, 652. [CrossRef]

14. Di Maio, P. Towards a Metamodel to Support the Joint Optimization of Socio Technical Systems. *Systems* **2014**, *2*, 273–296. [CrossRef]

15. Ahlborg, H.; Ruiz-Mercado, I.; Molander, S.; Masera, O. Bringing Technology into Social-Ecological Systems Research—Motivations for a Socio-Technical-Ecological Systems Approach. *Sustainability* **2019**, *11*, 2009. [CrossRef]

16. Kendall, E.; Oh, S.; Amsters, D.; Whitehead, M.; Hua, J.; Robinson, P.; Palipana, D.; Gall, A.; Cheung, M.; Potter, L.E.; et al. HabITec: A Sociotechnical Space for Promoting the Application of Technology to Rehabilitation. *Societies* **2019**, *9*, 74. [CrossRef]

17. Chacón, A.; Angulo, C.; Ponsa, P. Developing Cognitive Advisor Agents for Operators in Industry 4.0. In *New Trends in the Use of Artificial Intelligence for the Industry 4.0*; Martínez, L.R., Rios, R.A.O., Prieto, M.D., Eds.; IntechOpen: Rijeka, Croatia, 2020; Chapter 7. [CrossRef]

Article

Using Gaussian Mixture Models for Gesture Recognition During Haptically Guided Telemanipulation

Carlos J. Pérez-del-Pulgar [1,*], **Jan Smisek** [2,3], **Irene Rivas-Blanco** [1], **Andre Schiele** [2,3] and **Victor F. Muñoz** [1]

1   Department of Systems Engineering and Automation, Universidad de Málaga, Andalucía Tech, 29071 Málaga, Spain
2   Delft University of Technology, 2628 Delft, The Netherlands
3   Telerobotics and Haptics lab, European Space Agency, 2201 Noordwijk, The Netherlands
*   Correspondence: carlosperez@uma.es; Tel.: +34-951-952-324

Received: 14 June 2019; Accepted: 4 July 2019; Published: 10 July 2019

Abstract: Haptic guidance is a promising method for assisting an operator in solving robotic remote operation tasks. It can be implemented through different methods, such as virtual fixtures, where a predefined trajectory is used to generate guidance forces, or interactive guidance, where sensor measurements are used to assist the operator in real-time. During the last years, the use of learning from demonstration (LfD) has been proposed to perform interactive guidance based on simple tasks that are usually composed of a single stage. However, it would be desirable to improve this approach to solve complex tasks composed of several stages or gestures. This paper extends the LfD approach for object telemanipulation where the task to be solved is divided into a set of gestures that need to be detected. Thus, each gesture is previously trained and encoded within a Gaussian mixture model using LfD, and stored in a gesture library. During telemanipulation, depending on the sensory information, the gesture that is being carried out is recognized using the same LfD trained model for haptic guidance. The method was experimentally verified in a teleoperated peg-in-hole insertion task. A KUKA LWR4+ lightweight robot was remotely controlled with a Sigma.7 haptic device with LfD-based shared control. Finally, a comparison was carried out to evaluate the performance of Gaussian mixture models with a well-established gesture recognition method, continuous hidden Markov models, for the same task. Results show that the Gaussian mixture models (GMM)-based method slightly improves the success rate, with lower training and recognition processing times.

Keywords: robotics; telemanipulation; haptics; machine learning; gesture recognition

---

1. Introduction

In telemanipulation, a human operator performs a task in a distant environment by remotely controlling a robot. To allow efficient operation, the operator needs to receive sensory information from the remote site. Depending on the received information, telemanipulation can be classified as "direct" or "uni-lateral" [1], where there is no feedback to the operator, or "bilateral" [2], which enables dual interaction between the haptic and the operator. Although telemanipulation allows real-time human remote control, it is still considered to entail a rather high workload [3], at least compared with more supervisory or autonomous modes of operation. However, only telemanipulation allows reacting to unknown and unforeseen situations with spontaneous feedback. Therefore, enriching telemanipulation with additional automatic assistance would allow humans to perform complex tasks more efficiently.

In this sense, haptically guided telemanipulation [4] is shown to be a promising method that reduces the operator workload and can improve his or her performance. Haptic guidance is usually implemented by adding virtual channels into the feedback path (e.g., the force output of a virtual spring) that generates appropriate forces to constrain the operator input along pre-described reference trajectories. This method referred to in the literature as virtual fixtures [5] or active constraints [6]. Increased precision and safety, as well as a reduction in task completion time, is the promise of this control method. It has been applied to many different fields, such as remote assembly [7], telesurgery [8], vehicle control [9], and even space-to-ground telemanipulation with long time-delay [10]. To provide effective guidance feedback, reliable and accurate task position information is required, along with the trajectories to guide the operator. This is often obtained a priori from images or markers [5]. However, this approach entails many problems in real environments, where reference positions or trajectories are often affected by measurement errors [11] or even entirely unknown during complex manipulation. For instance, during insertion operations, virtual fixtures can hardly help if the guidance system does not know exactly the insertion point (its position and orientation), which is difficult to be obtained with any vision system due to occlusions and point-of-view limitations. To mitigate this problem, van Oosterhout et al. [12] suggested combining force feedback (robot interaction data) with guidance forces. However, how to derive meaningful, accurate and sufficiently well-computed guidance trajectories for real-time manipulation such that they augment natural human manipulation is still to be resolved. This problem was addressed in our previous contribution [13], which proposed to use a learning from demonstration (LfD) approach to provide real-time haptic guidance based on the use of interaction forces and torques. This method was successfully tested with the peg-in-hole insertion task, which is a de facto benchmark test for robotics assembly [14].

The main limitation of the previous approach is that the haptic assistance should be made dependent on the kind of movements the operator is performing at a given moment. Thus, the guidance trajectories need to be generated on the fly. For example, a dashboard panel could contain different switches and connectors. Depending on the task that is being carried out by the operator, e.g., operating an on/off switch or inserting a connector, different guidance references should be applied to solve the task. In this sense, Havoutis et al. [15] proposed to create a library of previously trained models that were used to complete each defined task autonomously. However, this contribution did not take into consideration any task recognition method.

It can be assumed that a simple task is equivalent to a gesture, and a complex task consists of several gestures. In this sense, gesture recognition has been widely studied with different methods and applications [16]. For years, gesture recognition has usually been addressed using a continuous or discrete hidden Markov model (HMM) [17,18]. An HMM can encode previously trained gestures as a sequence of states with probabilistic relationships between states and measurements. It has been commonly used to detect gestures once they have finished using the forward-backward algorithm [19], i.e., for a peg insertion task, the gesture would only be detected once the operator has already inserted the peg.

In this sense, learning from demonstration (LfD) is an approach that has been widely used to generate temporally continuous trajectories by teaching, based on the robot position or interaction. Indeed, LfD uses Gaussian mixture models (GMM) or continuous hidden Markov model (CHMM) to encode training trajectories and generate the most likely trajectory through Gaussian mixture regression (GMR). This approach allows robots to perform previously trained simple human tasks such as pouring a glass of water using a bimanual robot [20], hit a table tennis ball or feed a robotic doll [21], all of them using position references. Kronander et al. [22] proposed the use of the robot pose and the exerted forces to perform automated insertion tasks based on LfD. Moreover, recent contributions used LfD for different purposes such as learning robot-collaboration skills [23], performing automated tasks of underwater remotely operated vehicles [15] or doing housework autonomously [24]. In the field of haptically guided telemanipulation, LfD has been recently used to address different tasks related to surgical robotics. Chowriappa et al. [8] used LfD to optimize the trocar placement in minimally invasive surgery (MIS). They collected a set of forces, torques, and trajectories from multiple demonstrations of the task and encoded them through the LfD approach. Then, a generalization of this set of trajectories with its associated parameters was generated using Gaussian mixture regression (GMR). The trajectory was used to perform haptic guidance through virtual fixtures. This approach was experimentally tested in laparoscopic surgery, where the excessive load on the environment has to be avoided during the trocar insertion. Power et al. [17] proposed a LfD based framework for the position-based haptic guidance in surgical telemanipulation using gesture recognition. In this case, they used a CHMM to encode previously trained gestures (called primitive movements in the paper). This was used to solve different surgical tasks, such as needle-passing or peg-transfer. The model enabled recognition of the gesture that was performed and provided a suitable haptic guidance to the operator through virtual fixtures. However, this contribution only took the absolute instrument tip position as the measurement to detect the gesture, without taking into considerations any interaction measurement as forces or torques.

Summarizing, haptically guided telemanipulation based on interaction forces and torques solves the limitations of the methods that rely on predefined position-based trajectories. This approach has been proposed in our previous work [13] using an LfD based method. However, the best method to recognize the gesture that is being carried out, using the same model to generate the haptic guidance, remains to be investigated, e.g., some authors used GMM or CHMM to perform LfD without taking into consideration its performance for gesture recognition and vice versa. Therefore, this work is focused on a gesture recognition method based on the defined LfD approach, i.e., the use of GMM and/or CHMM. Thus, a complex task is divided into a set of simple gestures. Then, during the training stage, a GMM is encoded for each gesture and stored in a library. Hence, the system would be able to detect the gesture that is being carried out and provide a customized haptic assistance depending on the task the user is performing. Force-based gesture recognition has the additional advantage that it can be used for insertion manipulations, where position changes are hardly perceivable if some parts of a robot and an environment are in contact. Thus, we hypothesize that, for insertion and object assembly type of manipulations, a desirable guidance system should not encourage following a fixed time or position based trajectory. Furthermore, a criterion to evaluate how well each gesture has been trained to be recognized is proposed. A comparison between GMM and CHMM was carried out regarding CPU processing time and recognition accuracy. The feasibility was demonstrated in an end-to-end telemanipulation experiment in which several gestures related to the peg-in-hole insertion task were trained and recognized.

Briefly, the main contributions of this paper are, on the one hand, the use of the LfD approach to perform gesture detection, comparing the use of GMM instead of CHMM, and, on the other hand, a criterion, called GMM gesture detection score (GGDS), that can be used to choose the best number of Gaussians in a GMM, and analyze the difference between the trained gestures.

The paper is structured as follows. The proposed LfD method and the gesture recognition criteria are described in Section 2. Section 3 describes the reference task and shows the obtained experimental results. A comparison between CHMM and GMM is carried out in Section 4. Section 5 discusses the obtained results. Finally, conclusions and future works are reported in Section 6.

2. LfD for Gesture Recognition

Any complex "principle" task (such as inserting a peg in a hole) can be divided into a set of gestures $\Omega = \rho_1, ..., \rho_p$ (e.g., approach, make contact, adjust peg rotation based on force constraints, move along linear constraint, move up to rigid contact, etc.). Depending on the actual gesture, the required haptic guidance reference may be different (e.g., to align a peg, torques are predominantly used, whereas, to linearly guide during insertion, linear forces would be required). To allow such gesture-based feedback, there is a training stage that encodes the demonstrations of the operator into a library containing the set of gestures models, and a reproduction stage that recognizes the current task and provides the corresponding haptic guidance to the operator. Figure 1 shows the training stage, which is performed offline in a previous phase. During it, each of the gestures ρ_i is demonstrated u times using a training platform, e.g., a manipulator with kinesthetic movements, a teleoperated device, or a sensorized tool. The training device provides the interaction measurements, usually forces $\vec{f} = (f_x, f_y, f_z)$ and/or torques $\vec{\tau} = (\tau_x, \tau_y, \tau_z)$. These measurements are encoded as a Gaussian Mixture Model (GMM-GGDS), the fitness of which is evaluated before storing it within the Library Ω. Then, this library is used in the reproduction stage, shown in Figure 2, to provide haptic guidance during the teleoperation of a slave robot using a master haptic device. Interaction of the slave robot with the environment generates forces $\vec{f}$ and torques $\vec{\tau}$ that are obtained from the robot sensors. They are used by the gesture recognizer to output the gesture ρ_i that is being carried, and it is used to provide the haptic guidance reference to the operator using the method described in the previous contribution [18]. Hereafter, details of the gesture recognizer system are described. Table 1 summarizes the notation used to describe the proposed framework.

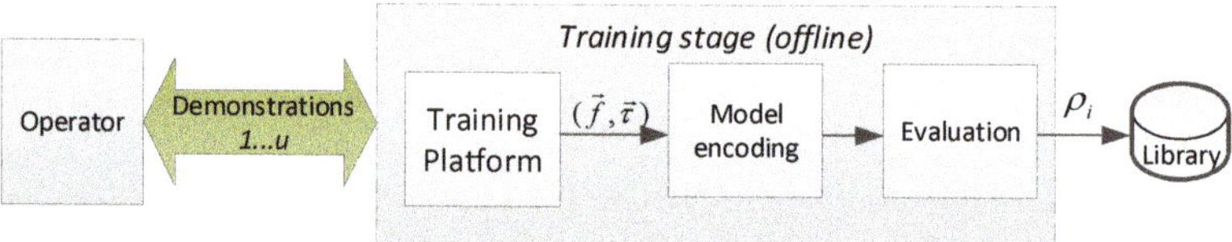

Figure 1. Training stage for gesture-based haptic guidance assistance. The operator performs a set of demonstrations for each gesture using a training platform. Then, each gesture is encoded as a Gaussian mixture models (GMM), evaluated and stored within the library Ω.

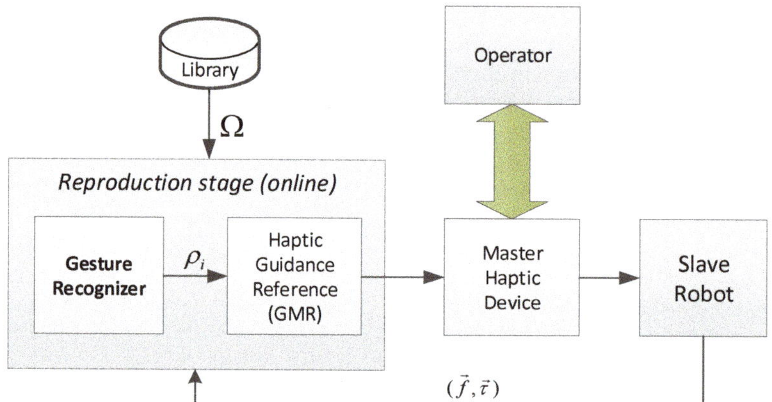

Figure 2. Method for gesture recognition and haptic guidance based on learning from demonstration. A unique library is used to detect the current gesture and provide the haptic guidance to the operator.

Table 1. Notation.

Symbol	Description
ρ_i	Encoded i gesture
Ω	Set of encoded gestures
$\vec{f}$	Exerted forces
$\vec{\tau}$	Exerted torques
ς	Training tuple
$\widehat{\varsigma}(k)$	Training sequence of k tuples

2.1. Training (Offline)

First, the operator demonstrates each gesture several times using a training device. During the u demonstration of the gesture i, a training sequence of k tuples is generated as:

$$\hat{\xi}_{iu}(k) = \xi(1), ..., \xi(k); 1 \leq i \leq p, 1 \leq u \leq U \tag{1}$$

where the training tuple is composed of the measured force and torque as:

$$\xi = (\vec{f}, \vec{\tau}) \tag{2}$$

Once the training sequences are obtained for each applicable gesture, they are used to encode a GMM ρ that will be used to recognize the gesture that is being carried out in real-time. A GMM is a probabilistic model that assumes the training sequences $\hat{\xi}_{i1...U}$ can be included in a set of N Gaussians distributions, whereas each distribution covers a part of the training sequences. Thus, a GMM can be defined as:

$$\rho = \{\pi_n, \mu_n, \Sigma_n\}_{n=1}^{N}, \tag{3}$$

with the following parameters:

- The number of Gaussians N is one of the most important parameters since this number affects the fitness and performance of the GMM. The gesture detection score, described in Section 2.2, is used to obtain this parameter.
- The prior probabilities π_n represent the weight of each Gaussian on the demonstrations, i.e., if a nth Gaussian covers more elements of the training sequences compared with another one, its prior probability will be higher.
- The means μ_n represent the centroid of each Gaussian of the GMM.
- The covariance matrices Σ_n define the amplitude of each Gaussian n.

The encoding of a GMM consists of adjusting the parameters of the GMM such that they fit with the training sequences. In this work, it is solved using the expectation-maximization (EM) algorithm [25]. The EM is an iterative method that can approximate the Gaussians to the training sequences, maximizing the likelihood of the training sequences belonging to the encoded GMM. It has been chosen for this purpose because it provides good results, in terms of accuracy and processing time, with low dimensions in the data. It is implemented as a function EM in Equation (4), whose input parameters are the training sequences for the gesture i: $\hat{\xi}_{i1...u}$ with u the number of demonstrations, and N the number of Gaussians.

$$\rho_i = \text{EM}(\hat{\xi}_{i1...u}, N) \tag{4}$$

The algorithm initializes each Gaussian with random parameters and they are adjusted to the training sequence iteratively as follows:

Expectation step:

$$\mathcal{P}(n|\xi_{i1...u}(j))^r = \frac{\pi_n^r \mathcal{N}(\xi_{i1...u}(j); \mu_n^r, \Sigma_n^r)}{\sum_{i=1}^{N} \pi_i^r \mathcal{N}(\xi_{i1...u}(j); \mu_i^r, \Sigma_i^r)} \tag{5}$$

$$\mathcal{E}_n^r = \sum_{j=1}^{k} \mathcal{P}(n|\xi_{i1...u}(j)^r) \tag{6}$$

Maximization step:

$$\pi_n^{r+1} = \frac{\mathcal{E}_n^r}{k} \tag{7}$$

$$\mu_n^{r+1} = \frac{\sum_{j=1}^{k} \mathcal{P}(n|\xi_{i1...u}(j))^r \xi_{i1...u}(j)}{\mathcal{E}_n^r} \tag{8}$$

$$\Sigma_n^{r+1} = \frac{\sum_{j=1}^{k} \mathcal{P}(n|\xi_{i1\ldots u}(j))^r (\xi_{i1\ldots u}(j) - \mu_n^{r+1})(\xi_{i1\ldots u}(j) - \mu_n^{r+1})^T}{\mathcal{E}_n^r} \tag{9}$$

In these equations, $\xi_{i1\ldots u}(j)$ is the tuple that corresponding to the position j on the training sequences for the gesture with k tuples, and r is the number of iterations. The iteration ends when the difference in log-likelihoods between the iterations is less than a predefined threshold C:

$$\frac{\ell(\widehat{\xi}_{i1\ldots u}|\rho^{r+1})}{\ell(\widehat{\xi}_{i1\ldots u}|\rho^r)} < C \tag{10}$$

2.2. Gesture Detection Score

As explained above, the number of Gaussians is an important parameter to be taken into account. Therefore, it may be selected based on the real-time constraint and evaluating how well every GMM fits its demonstrations. To perform this evaluation, the Bayesian information criterion (BIC) was previously used [13]. This method provides a score for different estimated GMMs, and it allows performing a comparison between them, taking the model fitness and dimension into account. However, in this paper, the trained GMMs is also used to recognize the gesture that is being carried out, and the BIC only provides information about the performance of different models for the same "gesture". It can be assumed separate gestures would be trained correctly using the BIC criteria. However, BIC does not provide information about how similar the gestures are. Thus, a GMM Gesture detection score (GGDS) has been defined in Equation (11) to overcome this issue. The GGDS provides a score for each gesture ρ_i that is calculated from the minimum difference between the log-likelihood of detecting other gestures ρ_j and the log-likelihood of detecting the gesture that is being evaluated ρ_i, using the training sequences obtained from the gesture i. The score provides information about how well the gesture will be correctly and incorrectly detected, compared with the rest of encoded gestures, with a lower score signifying a better model fitness.

$$GGDS_{\rho_i} = \min_{\rho_j \in \Omega, j \neq i} \left\{ \ell(\widehat{\xi}_{i1\ldots u}|\rho_j) - \ell(\widehat{\xi}_{i1\ldots u}|\rho_i) \right\} \tag{11}$$

The term $\ell(\widehat{\xi}_{i1\ldots u}|\rho)$ represents the log-likelihood that the training sequences belong to ρ and it is calculated as:

$$\ell(\widehat{\xi}_{i1\ldots u}|\rho) = \sum_{j=1}^{k} \log(\mathcal{P}(\xi(j)|\rho)) \tag{12}$$

with $\mathcal{P}(\xi(j)|\rho)$, the probability that the tuple $\xi(j)$ belongs to ρ, which can be calculated as:

$$\mathcal{P}(\xi|\rho) = \sum_{n=1}^{N} \pi_n \times \mathcal{N}(\xi; \mu_n, \Sigma_n) \tag{13}$$

where $\mathcal{N}(\xi; \mu_n, \Sigma_n)$ is the probability density function of ξ.

The GGDS provides a score that is useful to decide between different parametric GMMs taking into account they will be used to detect the gesture that is being carried out. Taking into account that the parameter to be optimized is the number of Gaussians N, several sets of gestures should be encoded using a different number of Gaussians, i.e., with $N = 1$ to $N = N_{MAX}$. For this purpose, Algorithm 1 performs training of p gestures for different number of Gaussians as follows. Firstly, every gesture is trained using the number of Gaussian (ranging from 1 to N_{MAX}) and the maximum GGDS score, comparing the gesture ρ_i with the rest of gestures, is stored in $SCORE_n$. Thus, the final number of Gaussians N is obtained from the minimum $SCORE_n$ for all the number of Gaussians. Finally, the trained gestures with the defined N are stored in the gesture library Ω.

Algorithm 1: Algorithm that obtains the best number of Gaussians N for the training sequences of several gestures using the GGDS criteria.

Data: $N_{MAX}, p, \widehat{\zeta}_{iu}(k)$
Result: N Optimal number of Gaussians to train every gesture
for $n = 1$ to N_{MAX} **do**
    for $i = 1$ to p **do**
        $\rho_i = EM(\widehat{\zeta}_{i1...u}, n)$
    end
    $SCORE_n = \max\limits_{\rho_i \in \Omega} \left[GGDS_{\rho_i} \right]$
end
$N = \underset{1 \leq n \leq N_{MAX}}{\text{argmin}} \left[SCORE_n \right]$
for $i = 1$ to p **do**
    $\rho_i = EM(\widehat{\zeta}_{i1...u}, N)$
    store ρ_i in Ω
end

2.3. Gesture Recognition (Online)

As stated before, each gesture was encoded into a GMM ρ in Equation (3) using training sequences that are composed of tuples ζ in Equation (2) that represent the interaction measurements. During the reproduction stage, the interaction measurements ζ are obtained each instant time from the robot sensors. Thus, the log-likelihood that the interaction measurements ζ belong to a GMM ρ can be expressed as: $log(\mathcal{P}(\zeta|\rho))$, which can be calculated from Equation (13). Thus, the most likely gesture i that is being carried out can be obtained from the gesture library as:

$$i = \underset{1 \leq i \leq p}{\text{argmax}} \left[log(\mathcal{P}(\zeta|\rho_i)) \right] \tag{14}$$

Analyzing these equations, it should be noted that the computational complexity of this method is $O(pN)$, which increases linearly with the number of Gaussians N and the number of gestures p. If N is too high, two cases can occur: (1) the processing time becomes too long for the real-time requirements during the haptic guidance owing to the large number of Gaussians in the GMM function; and/or (2) the improvement in the fitness of the GMM with respect to the training sequences is too low.

Once the gesture ρ_i in Equation (3) that is being carried out has been recognized, the corresponding model can be used to provide haptic guidance [13].

2.4. Reference Task: Peg-in-Hole Insertion with Tight Tolerance

As a de facto standard benchmark test for robotics assembly, the peg-in-hole insertion task was chosen to perform the experimental evaluation [14].

The peg-in-hole insertion task, despite being rather trivial when performed manually, has proven to be relatively challenging by the use of a robot (either teleoperated or performed autonomously) [26], in particular for long insertion dimension and tight, sub-millimeter tolerances.

As illustrated in Figure 3, we divided the peg-in-hole task into two gestures: surface contact (ρ_1) and lever effect (ρ_2), which depend on the actual interactions between the peg and hole during the execution of the task. At the beginning, during the first gesture, the operator attempts to position the peg at the entrance of the hole. This gesture is completed with establishing surface contact, still with negligence on correct orientation. Only in a second step, the operator will adjust alignment and guide the peg into the free direction for insertion.

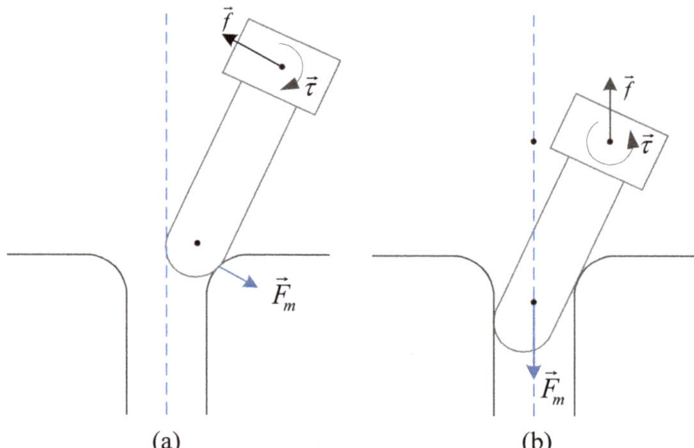

Figure 3. Peg-in-hole identified gestures: (**a**) surface contact gesture, the peg is touching the surface of the hole with lateral forces; and (**b**) lever effect gesture, the peg is already inserted in the hole without a correct orientation. Blue arrows represent the exerted forces.

In Figure 3a, the *surface contact* gesture is illustrated. The surface is pushed with a lateral force ($\vec{F}_m$). Because the peg is rigid, a reaction force of the same magnitude is transmitted to the base of the peg as $\vec{f}$. Furthermore, a small torque $\vec{\tau}$ is generated on the peg. In this case, the peg tip has to be moved horizontally to coincide with the hole. On the other hand, if the peg is already at the entrance of the hole without a correct orientation (in double contact, Figure 3b), the operator has to align the peg with the hole. In this situation, if the peg were "pushed" down with a force $\vec{F}_m$, vertical and opposite lateral forces $\vec{f}$ would arise because the peg is locked in the hole, and torques $\vec{\tau}$ occur in the opposite direction because of the lever effect. Thus, the peg may be rotated to align it with the hole. This is the *lever effect* gesture. To summarize, $\vec{f}$ and $\vec{\tau}$ represent the interaction measurements in which different magnitudes and directions are expected depending on the gesture that is being carried out. Once the interaction measurements for both gestures were obtained, the training tuple was defined as:

$$\xi = (f_x, f_y, \tau_x, \tau_y) \tag{15}$$

where f_x, f_y, τ_x and τ_y were obtained from a F/T sensor placed on the peg base. The parameters f_z and τ_z were removed from the tuple because they did not provide any relevant information to perform the guidance (only around peg symmetry axis).

3. Experimental Results

The peg-in-hole insertion task was experimentally validated by means of a telemanipulation platform located in the Telerobotics and Haptics Laboratory at the European Space Agency research center ESTEC.

The experimental validation was divided in two stages. The first one was related to the evaluation of the training stage, where two gesture were trained. The second one demonstrated the proposed gesture recognition approach and its validity for application in haptic guidance, where it should help operators to solve a task.

3.1. Experimental Setup

The experiments were carried out using a robotic telemanipulation workcell that is composed of a KUKA Lightweight Robot (LWR 4+) slave robot [27] and a Sigma.7 haptic master device [28] (Figure 4). Both devices were connected to the main computer that executed the proposed method, via the FRI Interface and UDP respectively. Figure 5 provides an experiment setup architectural overview. The taskboard, on the right-hand side of the figure, was used to research on various generic manipulation tasks. It contains different holes, pegs, and doors that can be operated on and used to validate various methods of telemanipulation and supervised autonomy. Here, the interaction with the taskboard was carried out by the manipulator Kuka LWR. During the telemanipulation, the manipulator was configured in impedance control mode to avoid any damage because of the interaction between the peg and the taskboard. During the training, the manipulator was configured in the gravity-compensated mode to allow an operator to move the robot by hand and position it freely in space to perform the training stage (hence, teach) certain task elements. This feature was used during the training stage to perform the peg-in-hole insertions by human-guided kinesthetic movements in a more direct way (as opposed to using a master device remotely for this stage).

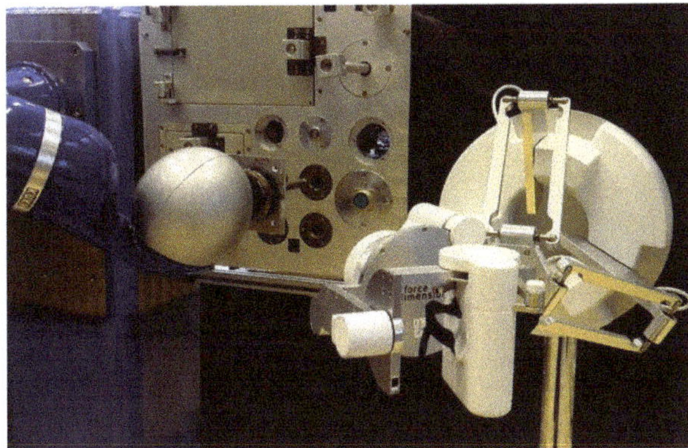

Figure 4. Overview of the Experimental setup used for learning from demonstration (LfD)-based haptic guidance, with the master device (Sigma.7 haptic master) and the KUKA lightweight slave robot manipulating a metal peg into a task-board receptacle.

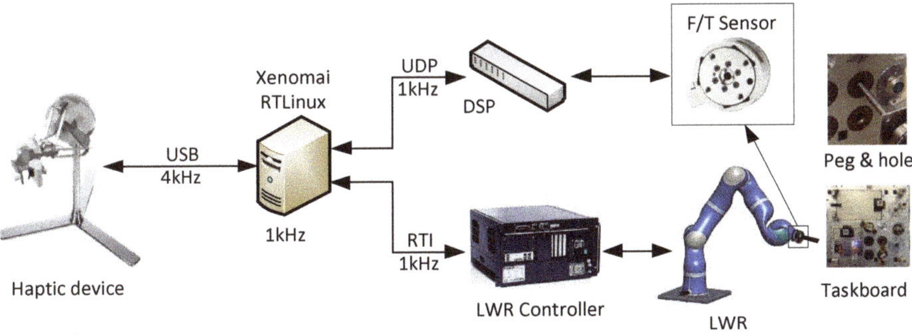

Figure 5. Software and Hardware architecture of the experimental telemanipulation setup as available at the Telerobotics Laboratory of the European Space Agency.

A 155 mm long titanium peg (approximate mechanical circumferential tolerance of 150 micrometers to the diameter of the metal receptacle in task-board) was rigidly mounted on a multiaxial Force/Torque (F/T) sensor (ATI Net F/T Gamma SI 65-5) attached to the robot's end-effector. The sensor measures interaction forces $\vec{f} = (f_x, f_y, f_z)$ and torques $\vec{\tau} = (\tau_x, \tau_y, \tau_z)$ on the peg base, providing the necessary information to the method described above. During task-reproduction, the manipulator was teleoperated by a force dimension Sigma.7 haptic device. It was used to receive position references from the operator, but also provided force feedback to the operator to guide him to perform the detected gesture. Forces were scaled to resist another operator movement during the assisted teleoperation. All of these devices, except the haptic one, which was connected directly through USB, were communicating with the main computer using the RTI Data Distribution Service middleware over a dedicated gigabyte LAN network. The main computer was a PC with a real-time Linux-based operating system (Xenomai). The computer algorithms were implemented with 1 ms sampling time.

3.2. Training

To obtain the training sequences for both gestures, the LWR was manually guided with its peg into the hole from eight different and $90°$ offset starting positions, which varied for the two types of gestures. Figure 6a–d presents the defined initial positions that cover a circle. Figure 6e,f presents the starting positions of the peg for the lever effect and the surface contact gesture, respectively. Three movements were performed for each gesture and each initial position, resulting in 12 insertion movements. They were used to train a model for each gesture. On the other hand, another 12 insertion movements were carried out using the complete telemanipulation workcell, which were used to evaluate the proposed approach.

The training insertions were used to encode both GMMs with a different number of Gaussians N, and they were evaluated through BIC and GGDS using Equation (11). Figure 7 depicts a comparison of the obtained scores between the different number of Gaussians for both gestures and scores. It demonstrates how the GGDS score was improved exponentially as the number of Gaussians is increased (larger negative score). However, the score obtained using the BIC increased almost linearly, because it does not provide information about how well the gesture would be recognized with respect the rest. Moreover, GGDS indicates both gestures would be recognized similarly as both of them received a similar score for an increased number of Gaussians. As stated before, the reproduction stage processing time increases linearly with the number of Gaussians. A test was performed to determine the maximum number of Gaussians that the main computer was able to execute within the real-time constraints of the system used in telemanipulation (characterized by a 1 ms sampling time). For this purpose, each gesture was trained with the number of Gaussians ranging from 1 to 20. Later on, the recognition algorithm was executed in the computer for each trained model to evaluate if the computer was able to execute it within the 1 ms constraint. Finally, $N = 14$ was chosen to carry out the experiments on the computer platform.

The EM algorithm in Equation (4) was used to train a GMM for each gesture. Figure 8a,b illustrates the training sequences and the Gaussians for both gestures. The training sequences $\hat{\zeta}_{i1...U}$ are represented by colored dots, one color for each training sequence. Each dot represents interaction measurement tuples ζ^m obtained at every time instant. For simplicity, only forces f_x and f_y are shown for the surface contact gesture and torques τ_x and τ_y for the lever effect gesture. The reason for choosing these measurements is because they are the most relevant information for each gesture, as stated above. Finally, the encoded Gaussians are represented by crosses (means) and ellipses (covariances). Once the best-encoded GMMs were selected, they were stored in the gesture library.

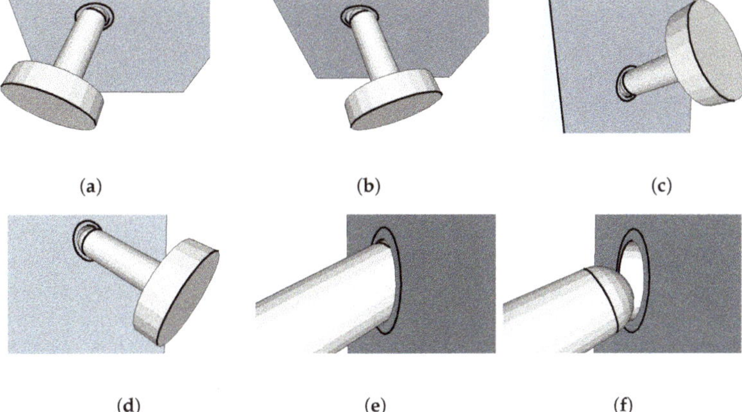

(a) **(b)** **(c)**

(d) **(e)** **(f)**

Figure 6. Initial training positions: (**a–d**) four different insertion positions to cover a circle over the hole; (**e**) initial position of the surface contact gesture; and (**f**) initial position of the lever effect gesture. (**a**) Left; (**b**) right; (**c**) up; (**d**) down; (**e**) lever effect; and (**f**) surface contact are shown.

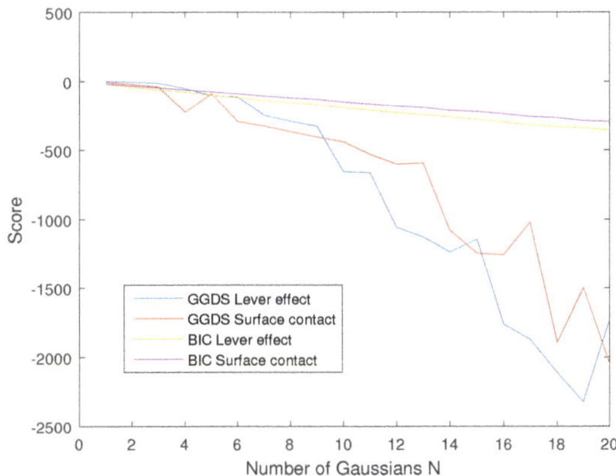

Figure 7. GMM gesture detection score (GGDS) for both gestures using different number of Gaussians. The figure shows the GGDS score decreased exponentially as the number of Gaussians was increased.

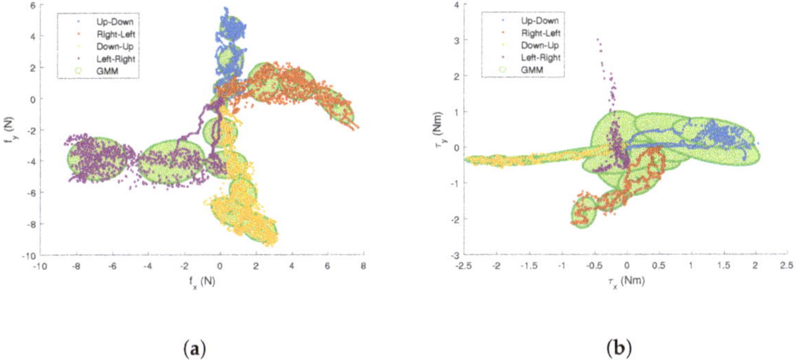

(a) **(b)**

Figure 8. GMM encoding for the surface contact (**a**) and lever effect (**b**) gestures showing the fit of the training sequences with the generated Gaussians. The colored points represent the training measurements for each movement.

3.3. Reproduction

Once populated and evaluated, the gesture library was used during the reproduction stage to recognize the gesture that was being carried out online, i.e., at the very instant in time. Moreover, modified models stored in the gesture library could be used to generate the appropriate haptic guidance reference to assist an operator during the telemanipulation execution [13].

As stated above, the interaction measurements $\vec{f}$ and $\vec{\tau}$ were used to calculate the most likely gesture that was being carried out through Equation (14). All evaluation movements were used to validate the proposed method. Figure 9 illustrates how the surface contact gesture is detected. Here, four evaluation movements were used, each one starting from different initial positions (see Figure 6a–d). In Figure 9a–d, the interaction measurements, obtained at each instant time, are represented by colored lines, i.e., one color for each movement from the various starting points. The trained Gaussian means are represented by crosses and the covariances by the ellipses. These figures show how the selected motion sequences enabled the algorithms to recognize the surface contact gestures (Figure 9a,b) and disabled their ability to recognize the lever effect gesture (Figure 9c,d). Figure 9e represents the log-likelihood of each evaluation sequence that belongs to the surface contact gesture GMM and Figure 9f represents the log-likelihood for the lever effect gesture. As shown, the log-likelihood of the evaluation movements for the surface contact motions was mostly higher for the surface contact gesture (-9 to 6 in the figure) than for the lever effect one (-90 to -15), which indicates the gesture was correctly recognized.

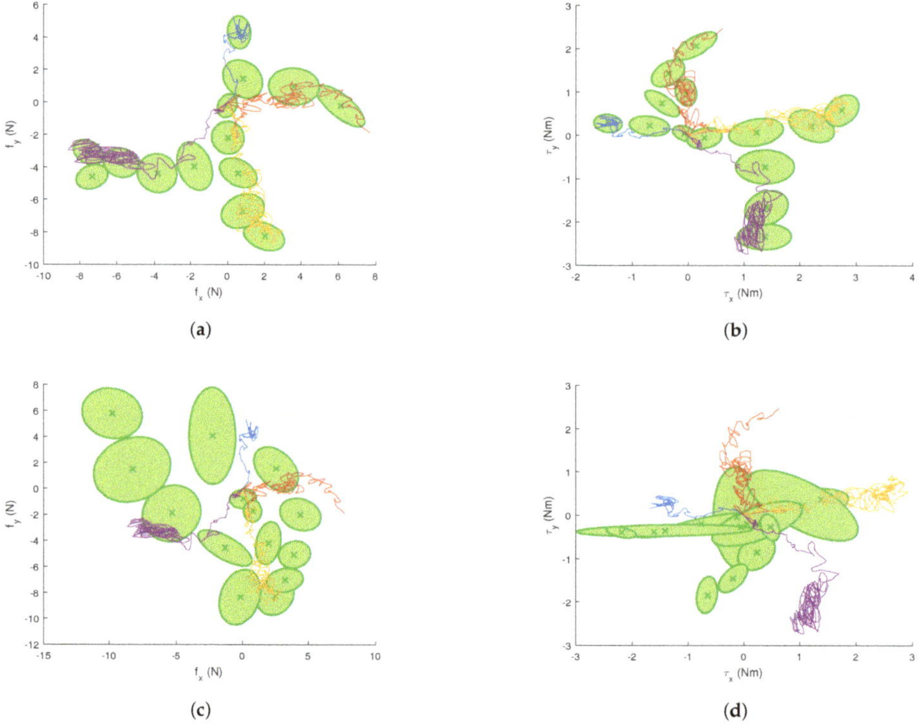

Figure 9. *Cont.*

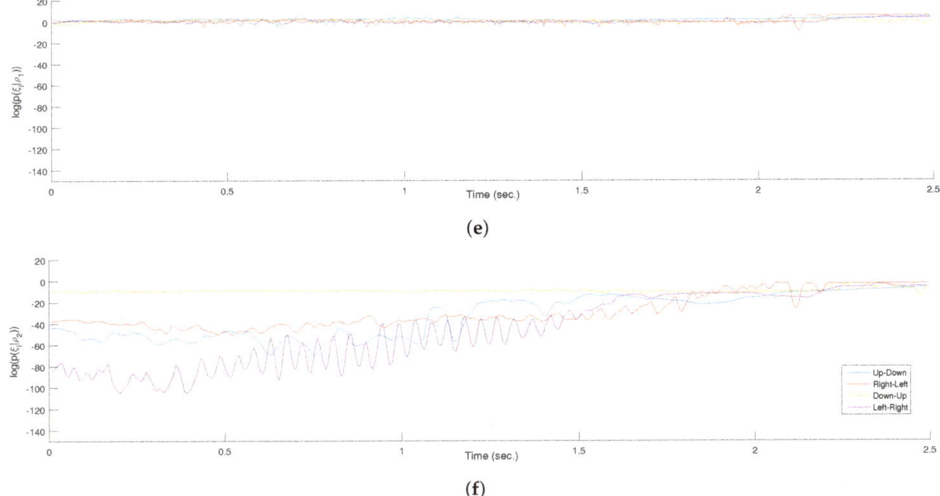

Figure 9. Trained surface contact (**a**) and lever effect (**b**) GMMs, and their fitnesses with the surface contact gesture interaction measurements from the evaluation movements. Log-likelihood of the surface contact gesture interaction measurements for both gestures (**c**,**d**). (**a**) Interaction forces vs. surface contact gesture GMM; (**b**) interaction torques vs. surface contact gesture GMM; (**c**) interaction forces vs. lever effect gesture GMM; (**d**) interaction torques vs. lever effect gesture GMM; (**e**) log-likelihood that the interaction measurement tuple ξ_i, obtained each i instant time, belongs to the surface contact GMM ρ_1; and (**f**) log-likelihood that the interaction measurement tuple ξ_i, obtained each i instant time, belongs to the lever effect GMM ρ_2.

4. Comparison between GMM and CHMM

Taking into consideration that CHMM has been widely used for gesture recognition, GMM arises as an alternative for haptically guided telemanipulation tasks. Indeed, the main advantages of this method are the lower computer complexity for training and reproduction. The CHMM computer complexity for gesture recognition is $O(PN^2T)$ where P is the number of gestures, N is the number of states, and T is the number of measurements. Contrarily, the computer complexity of the GMM based gesture recognition method is $O(PN)$, i.e., the processing time increases linearly. In the case of the training stage, CHMM uses the k-means algorithm to define each continuous hidden state, whose computer complexity is: $O(NTK)$, where K is the number of iterations. Afterwards, the Baum–Welch algorithm is used to train the HMM itself according to the previously defined hidden states. The computer complexity of this algorithm is $O(NT)$, resulting in $O(NTK) + O(NT)$ for the CHMM. However, GMM only uses the k-means algorithm to encode a gesture model.

To evaluate the performance of GMM versus CHMM experimentally, previous measurements for training and evaluation were used to encode and evaluate a CHMM with the same parameters as the previously encoded GMM. The results obtained for the CPU processing time are shown in Figure 10, where it can be seen that GMM achieved slightly better results than CHMM. As regards the recognition rate, each model was used to recognize the performed gesture that was being carried out in real-time, i.e., every sampling period (1 ms) the gesture was recognized. As shown in Figure 11, for the lever effect gesture, CHMM was slightly better than GMM, but, in the case of the surface contact one, GMM was better than CHMM.

On the other hand, the detection of the gesture can be seen as a multi-objective optimization problem, where there are a cost function and an objective one, i.e., the CPU processing time and the log-likelihood of the detected gesture. Therefore, the objective is to minimize the CPU processing time and maximize the log-likelihood. Figure 12 represents the cost and objective for both methods and gestures. Every evaluation movement was trained using a different number of Gaussians or states, obtaining different values as it was increased. The log-likelihood represented in this figure has been calculated according to Equation (12). Results show that GMM provides better results using the same CPU processing time than CHMM, which required approximately double the processing time to obtain equal log-likelihood for the same movement.

Finally, an evaluation of training the gestures using a different number of Gaussians or states was carried out. Figure 13 represents the CPU processing time to train both gestures using GMM and CHMM. As shown, the needed processing time for GMM was much lower than CHMM as the number of Gaussians was increased.

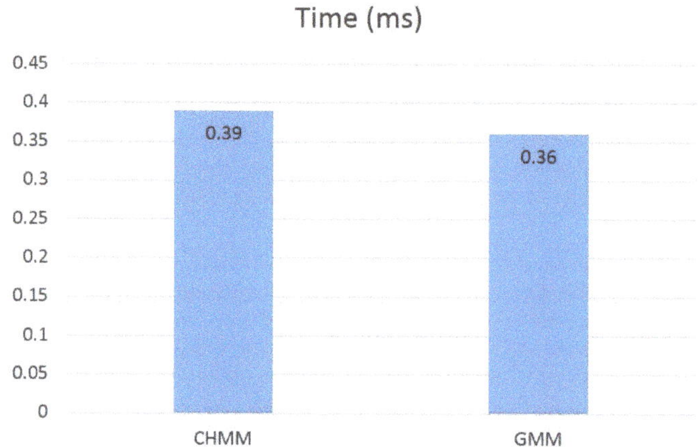

Figure 10. Average CPU processing time during the recognition of each evaluation movement using both continuous hidden Markov model (CHMM) and GMM methods.

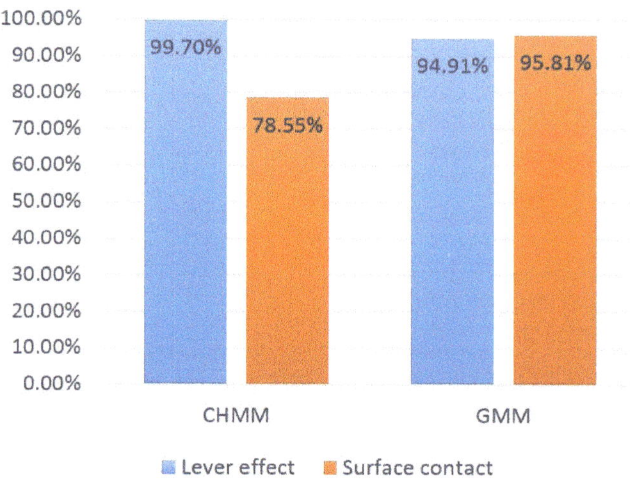

Figure 11. Percentage of real-time recognition using the evaluation movement of both gestures, and both CHMM and GMM methods.

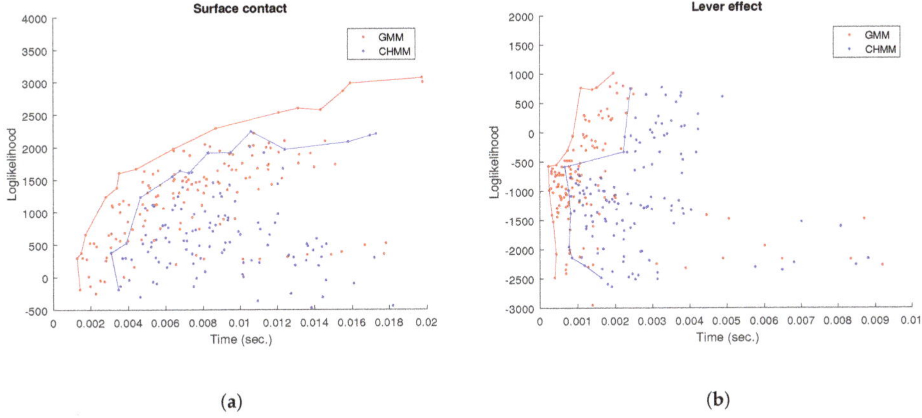

(**a**)   (**b**)

Figure 12. Pareto frontier for GMM and CHMM using the log-likelihood and processing time of each evaluation movement for both gestures: (**a**) surface contact gesture; and (**b**) lever effect gesture.

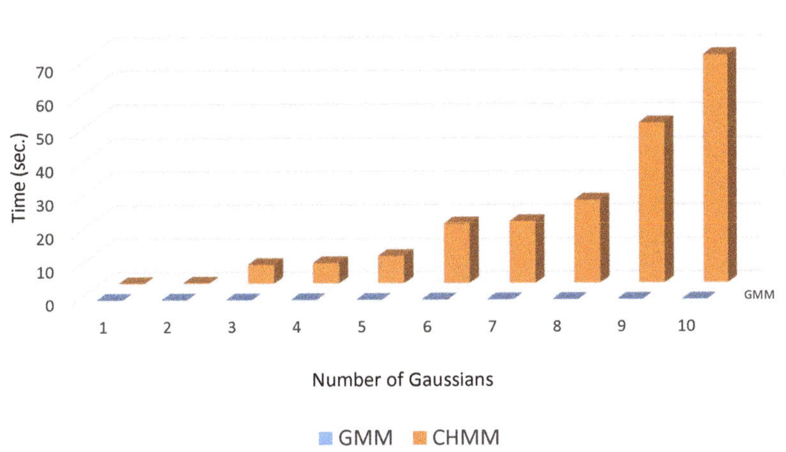

Figure 13. Training time using GMM and CHMM for different number of Gaussians/states.

5. Discussion

The above results show that it is feasible to recognize, in real-time, the gesture that is being carried out with a high success rate. The presented method was validated with the peg-in-hole insertion task under sub-millimeter tolerances. During the training stage, two gestures were defined through 24 demonstrations that generated two GMMs, composed of 14 Gaussians each. Although the presented work was focused on two manually predefined gestures (i.e., surface contact and lever effect) the proposed method could be used with recently proposed segmentation of movements techniques [29].

The defined GMM gesture detection score (GGDS) was used to analyze the GMM performance taking into account the number of Gaussians, which demonstrates that the performance gets better when the number of Gaussians is increased for the peg-in-hole insertion task. Therefore, the main limitation to choose the maximum number of Gaussians was the real-time constraint of 1 ms. However, this method has a limitation: as it has been defined, the same number of Gaussians is used to train every gesture, which would not optimize the right number of Gaussians if there are simple and complex gestures. It is worth mentioning the GGDS is also useful to detect how different the gestures to be trained are, which would provide information about two or more similar gestures. In this case, the GGDS would help to decide the best number of Gaussians.

During the gesture recognition, the GMM-based method achieved a success rate of 94.91% and 95.81% for each gesture. The CHMM-based method provided a success rate of 99.7% and 78.6% for the same training and evaluation sequences. Although the lever effect gesture was better recognized by CHMM, the recognition of the surface contact was significantly worse. The authors consider that CHMM is intended to detect the gesture once it has ended, and it is assumed that this is the reason because GMMs provide better results than CHMM during real-time gesture recognition. Moreover, another drawback of CHMM is that the method is able to detect a gesture by a motion sequence, due to the use of hidden states. However, GMMs detect the gesture for the motion data every instant time, i.e., an incorrect detection at the beginning of the motion would not affect the entire recognition of the gesture.

Safety of the method, in terms of gesture recognition accuracy, should be taken into consideration to avoid a wrong guidance. Although this paper does not provide any solution, a good approach would be to take into consideration a log-likelihood threshold to enable the haptic guidance for the detected gesture. Depending on it, the system would be more or less conservative providing haptic guidance.

On the other hand, a Pareto analysis was carried out, which shows that GMM provides better performance than CHMM during gesture detection, by providing comparable result in half the processing time. During the training stage, both models were trained with the number of Gaussians between 1 and 10. The required processing time for GMM ranged between 0.00257 s and 0.020997 s and between 0.141 s and 67.980 s for CHMM. Such a large difference in the training stage can be especially advantageous in certain applications, for example when used in future reinforcement learning methods.

6. Conclusions

This work shows the use of a method based on LfD for real-time gesture recognition during telemanipulation. The developed method performs sufficiently well to allow gesture recognition in real-time with a one-millisecond sampling-time constraint. Moreover, it is shown that a 14 Gaussians GMM approach is sufficient to adequately recognize the gestures of a peg-in-hole insertion task (that is here characterized by two individual gestures: "establishing of surface contact" and "levering the peg into the hole"). The new "interaction-force" based method proposed in this paper provides a viable way to encode and reconstruct contact tasks in robotic assembly to guide human operators whenever either visual information is limited or when positional or time-based indexed guidance data are not available or difficult to achieve. The LfD-based framework proposed by us earlier [18] for haptic shared control is extended by an LfD-based gesture recognition module, which allows application in real-time, due to its linear increase of computational complexity with the number of Gaussian Models used to encode any gesture. We consider the proposed approach to be extendable to other complex manipulation tasks involving contact operations, such as operation on knobs, doors, hinges, and levers, and, therefore, we assume that the method can significantly improve remotely operated robotic manipulation in a variety of contexts. Moreover, a further comparison among GMM, CHMM and other recognition methods would be useful to analyze suitability in specific situations. These issues are proposed to be addressed as future work.

Author Contributions: Conceptualization, C.J.P.-d.-P., J.S. and A.S.; Methodology, C.J.P.-d.-P., A.S. and V.F.M.; Software, C.J.P.-d.-P. and J.S.; Validation, C.J.P.-d.-P., J.S. and I.R.-B.; Writing—original draft preparation, C.J.P.-d.-P., I.R.-B.; Writing—review and editing, A.S., J.S., I.R.-B. and V.F.M.; and Supervision, A.S. and V.F.M.

Funding: This research was partially funded by Ministry of Science and Innovaction (Spain), under the project code DPI2016-80391-C3-1-R.

Acknowledgments: Authors would like to thank the Telerobotics and Haptic Laboratory at the European Space Agency for supporting this work.

Conflicts of Interest: The authors declare no conflict of interest.

Abbreviations

The following abbreviations are used in this manuscript:

LfD	Learning from Demonstration
HMM	Hidden Markov Model
GMM	Gaussian Mixture Models
GMR	Gaussian Mixture Regression
CHMM	Continuous Hidden Markov Model
MIS	Minimally Invasive Surgery
EM	Expectation-Maximization
BIC	Bayesian Information Criterion
GGDS	Gesture Detection Score

References

1. Codourey, A.; Rodriguez, M.; Pappas, I. A task-oriented teleoperation system for assembly in the microworld. In Proceedings of the 1997 8th International Conference on Advanced Robotics, ICAR'97, Monterey, CA, USA, 7–9 July 1997; IEEE: Piscataway, NJ, USA, 1997; pp. 235–240.
2. Hokayem, P.F.; Spong, M.W. Bilateral teleoperation: An historical survey. *Automatica* **2006**, *42*, 2035–2057. [CrossRef]
3. Chen, J.Y.; Haas, E.C.; Barnes, M.J. Human performance issues and user interface design for teleoperated robots. *IEEE Trans. Syst. Man Cybern. Part C Appl. Rev.* **2007**, *37*, 1231–1245. [CrossRef]
4. Bolopion, A.; Xie, H.; Haliyo, D.S.; Régnier, S. Haptic teleoperation for 3-D microassembly of spherical objects. *IEEE/ASME Trans. Mechatron.* **2012**, *17*, 116–127. [CrossRef]
5. Abbott, J.J.; Marayong, P.; Okamura, A.M. Haptic virtual fixtures for robot-assisted manipulation. In *Robotics Research*; Springer: Berlin, Germany, 2007; pp. 49–64.
6. Nuno, E.; Rodríguez, A.; Basañez, L. Force reflecting teleoperation via ipv6 protocol with geometric constraints haptic guidance. In *Advances in Telerobotics*; Springer: Berlin, Germany, 2007; pp. 445–458.
7. Boessenkool, H.; Abbink, D.A.; Heemskerk, C.J.; van der Helm, F.C.; Wildenbeest, J.G. A task-specific analysis of the benefit of haptic shared control during telemanipulation. *Haptics IEEE Trans.* **2013**, *6*, 2–12. [CrossRef] [PubMed]
8. Chowriappa, A.; Wirz, R.; Ashammagari, A.R.; Seo, Y.W. Prediction from expert demonstrations for safe tele-surgery. *Int. J. Autom. Comput.* **2013**, *10*, 487–497. [CrossRef]
9. Griffiths, P.G.; Gillespie, R.B. Sharing control between humans and automation using haptic interface: Primary and secondary task performance benefits. *Hum. Factors J. Hum. Factors Ergon. Soc.* **2005**, *47*, 574–590. [CrossRef] [PubMed]
10. Schiele, A. Towards the Interact Space Experiment: Controlling an Outdoor Robot on Earth's Surface from Space. In Proceedings of the 13th Symposium on Advanced Space Technologies for Robotics and Automation (ASTRA), Noordwijk, The Netherlands, 11–13 May 2015; European Space Agency: Paris, France, 2016.
11. Smisek, J.; van Paassen, M.; Schiele, A. Haptic guidance in bilateral teleoperation: Effects of guidance inaccuracy. In Proceedings of the 2015 IEEE World Haptics Conference (WHC), Evanston, IL, USA, 22–26 June 2015; IEEE: Piscataway, NJ, USA, 2015; pp. 500–505.
12. Van Oosterhout, J.; Wildenbeest, J.G.; Boessenkool, H.; Heemskerk, C.J.; de Baar, M.R.; van der Helm, F.C.; Abbink, D.A. Haptic Shared Control in Tele-Manipulation: Effects of Inaccuracies in Guidance on Task Execution. *IEEE Trans. Haptics* **2015**, *8*, 164–175. [CrossRef] [PubMed]
13. Pérez-del Pulgar, C.J.; Smisek, J.; Muñoz, V.F.; Schiele, A. Using learning from demonstration to generate real-time guidance for haptic shared control. In Proceedings of the 2016 IEEE International Conference on Systems, Man, and Cybernetics (SMC), Budapest, Hungary, 9–12 October 2016; IEEE: Piscataway, NJ, USA, 2016; pp. 003205–003210.

14. Unger, B.J.; Nicolaidis, A.; Berkelman, P.J.; Thompson, A.; Klatzky, R.L.; Hollis, R.L. Comparison of 3-D haptic peg-in-hole tasks in real and virtual environments. In Proceedings of the 2001 IEEE/RSJ International Conference on Intelligent Robots and Systems. Expanding the Societal Role of Robotics in the the Next Millennium (Cat. No.01CH37180), Maui, HI, USA, 29 October–3 November 2001; IEEE: Piscataway, NJ, USA, 2001; Volume 3, pp. 1751–1756.

15. Havoutis, I.; Calinon, S. Supervisory teleoperation with online learning and optimal control. In Proceedings of the 2017 IEEE International Conference on Robotics and Automation (ICRA), Singapore, 29 May–3 June 2017.

16. Mitra, S.; Acharya, T. Gesture recognition: A survey. *IEEE Trans. Syst. Man Cybern. Part C Appl. Rev.* **2007**, *37*, 311–324. [CrossRef]

17. Power, M.; Rafii-Tari, H.; Bergeles, C.; Vitiello, V.; Yang, G.Z. A cooperative control framework for haptic guidance of bimanual surgical tasks based on Learning From Demonstration. In Proceedings of the 2015 IEEE International Conference on Robotics and Automation (ICRA), Seattle, WA, USA, 26–30 May 2015; IEEE: Piscataway, NJ, USA, 2015; pp. 5330–5337.

18. Del Pulgar, C.J.P.; Garcia-Morales, I.; Blanco, I.R.; Munoz, V.F. Navigation Method for Teleoperated Single-Port Access Surgery With Soft Tissue Interaction Detection. *IEEE Syst. J.* **2016**, in press. [CrossRef]

19. Rabiner, L.R. A tutorial on hidden Markov models and selected applications in speech recognition. *Proc. IEEE* **1989**, *77*, 257–286. [CrossRef]

20. Jäkel, R.; Schmidt-Rohr, S.R.; Lösch, M.; Dillmann, R. Representation and constrained planning of manipulation strategies in the context of programming by demonstration. In Proceedings of the 2010 IEEE International Conference on Robotics and Automation, Anchorage, AK, USA, 3–7 May 2010; IEEE: Piscataway, NJ, USA, 2010; pp. 162–169.

21. Calinon, S.; D'halluin, F.; Sauser, E.L.; Caldwell, D.G.; Billard, A.G. Learning and reproduction of gestures by imitation. *IEEE Robot. Autom. Mag.* **2010**, *17*, 44–54. [CrossRef]

22. Kronander, K.; Burdet, E.; Billard, A. Task transfer via collaborative manipulation for insertion assembly. In *Workshop on Human-Robot Interaction for Industrial Manufacturing, Robotics, Science and Systems*; Citeseer: Centre County, PA, USA, 2014.

23. Rozo, L.; Calinon, S.; Caldwell, D.G.; Jimenez, P.; Torras, C. Learning physical collaborative robot behaviors from human demonstrations. *IEEE Trans. Robot.* **2016**, *32*, 513–527. [CrossRef]

24. Kormushev, P.; Calinon, S.; Caldwell, D.G. Imitation learning of positional and force skills demonstrated via kinesthetic teaching and haptic input. *Adv. Robot.* **2011**, *25*, 581–603. [CrossRef]

25. Bilmes, J.A. A gentle tutorial of the EM algorithm and its application to parameter estimation for Gaussian mixture and hidden Markov models. *Int. Comput. Sci. Inst.* **1998**, *4*, 126.

26. Chhatpar, S.R.; Branicky, M.S. Search strategies for peg-in-hole assemblies with position uncertainty. In Proceedings of the 2001 IEEE/RSJ International Conference on Intelligent Robots and Systems. Expanding the Societal Role of Robotics in the the Next Millennium (Cat. No.01CH37180), Maui, HI, USA, 29 October–3 November 2001; IEEE: Piscataway, NJ, USA, 2001; Volume 3, pp. 1465–1470.

27. Bischoff, R.; Kurth, J.; Schreiber, G.; Koeppe, R.; Albu-Schäffer, A.; Beyer, A.; Eiberger, O.; Haddadin, S.; Stemmer, A.; Grunwald, G.; et al. The KUKA-DLR Lightweight Robot arm-a new reference platform for robotics research and manufacturing. In Proceedings of the ISR 2010 (41st International Symposium on Robotics) and ROBOTIK 2010 (6th German Conference on Robotics), VDE, Munich, Germany, 7–9 June 2010; pp. 1–8.

28. Tobergte, A.; Helmer, P.; Hagn, U.; Rouiller, P.; Thielmann, S.; Grange, S.; Albu-Schäffer, A.; Conti, F.; Hirzinger, G. The sigma. 7 haptic interface for MiroSurge: A new bi-manual surgical console. In Proceedings of the 2011 IEEE/RSJ International Conference on Intelligent Robots and Systems, San Francisco, CA, USA, 25–30 September 2011; IEEE: Piscataway, NJ, USA, 2011; pp. 3023–3030.

29. Figueroa, N.; Ureche, A.L.P.; Billard, A. Learning complex sequential tasks from demonstration: A pizza dough rolling case study. In Proceedings of the 2016 11th ACM/IEEE International Conference on Human-Robot Interaction (HRI), Christchurch, New Zealand, 7–10 March 2016; IEEE: Piscataway, NJ, USA, 2016; pp. 611–612.

 *electronics*

Article

Transferring Know-How for an Autonomous Camera Robotic Assistant

Irene Rivas-Blanco *, Carlos J. Perez-del-Pulgar , Carmen López-Casado, Enrique Bauzano and Víctor F. Muñoz

Department of Systems Engineering and Automation, Universidad de Málaga, Andalucía Tech, 29071 Málaga, Spain; carlosperez@uma.es (C.J.P.-d.-P.); mclopezc@uma.es (C.L.-C.); ebauzano@uma.es (E.B.); vfmm@uma.es (V.F.M.)
* Correspondence: irivas@uma.es; Tel.: +34-952-137-204

Received: 10 January 2019; Accepted: 13 February 2019; Published: 18 February 2019

Abstract: Robotic platforms are taking their place in the operating room because they provide more stability and accuracy during surgery. Although most of these platforms are teleoperated, a lot of research is currently being carried out to design collaborative platforms. The objective is to reduce the surgeon workload through the automation of secondary or auxiliary tasks, which would benefit both surgeons and patients by facilitating the surgery and reducing the operation time. One of the most important secondary tasks is the endoscopic camera guidance, whose automation would allow the surgeon to be concentrated on handling the surgical instruments. This paper proposes a novel autonomous camera guidance approach for laparoscopic surgery. It is based on learning from demonstration (LfD), which has demonstrated its feasibility to transfer knowledge from humans to robots by means of multiple expert showings. The proposed approach has been validated using an experimental surgical robotic platform to perform peg transferring, a typical task that is used to train human skills in laparoscopic surgery. The results show that camera guidance can be easily trained by a surgeon for a particular task. Later, it can be autonomously reproduced in a similar way to one carried out by a human. Therefore, the results demonstrate that the use of learning from demonstration is a suitable method to perform autonomous camera guidance in collaborative surgical robotic platforms.

Keywords: surgical robotics; human–machine interaction; autonomous guidance

---

1. Introduction

In the last decades, surgical robotics has spread to the operating rooms as a daily reality. The Da Vinci surgical system (Intuitive Surgical, Inc.), the most used commercial surgical robot, has distributed more than 4000 units in hospitals around the world that have been used to perform more than five million procedures. This platform is composed of a slave side that replicates the motions, and a surgeon who performs in a master console [1]. Although this kind of robot enhances the surgeon's abilities, providing more stability and accuracy to the surgical instruments, their assistance is limited to imitating the movements performed on the master console. However, all the cognitive load lies on the surgeon, who has to perform every single motion of the endoscopic tools. The implementation of collaborative strategies to perform autonomous auxiliary tasks would reduce the surgeon's workload during the interventions, letting she or he concentrate on the important surgical task.

Camera guidance is a particularly interesting task to be automated, as it is a relatively simple but crucial task that may significantly help the surgeons. Most robotic surgeries are performed by a single surgeon who controls both the instruments and the camera, switching the control between them through a pedal interface. Omote et al. [2] addressed the use of a self-guided robotic laparoscopic camera based on a color tracking method to follow the instruments. In this work, they compared this method with human camera control and demonstrated that autonomous camera guidance slightly reduces the surgery time, and the camera corrections and lens cleaning frequency were drastically improved. Therefore, this approach helps the surgeon to be more concentrated on the surgery.

Current camera guidance strategies can be divided into three approaches; direct control interfaces, instruments tracking, and cognitive approaches. Direct control interfaces, such as voice commands [3], head movements [4], or gaze-contingent camera control [5], have succeeded in substituting medical staff for moving the camera. However, these methods require very specific instructions and extraneous devices that may distract the surgeon. On the other hand, camera guidance through instrument tracking gives the robot more autonomy and does not require continuous surgeon supervision [6–8]. The earliest works employed color markers to identify the tip of the tools, but the current trend is to use deep learning techniques to identify the tools in raw endoscopic images [9–11]. Reference [12] improves traditional tracking methods with the long-term prediction of the surgical tools motion using Markov chains.

The main problem of these strategies is that they follow simple and rigid rules, such as tracking only one tool, or the middle point of both tools [13], but they do not consider other important factors that affect the behavior of the camera such as the knowledge of what task the surgeon is performing at a given time or particular camera view preferences.

Hence, cognitive approaches emerge to provide more flexibility to the camera behavior, making the guidance strategy dependent on the current state of the task. Reference [14] proposes a human–robot interaction architecture that sets a particular camera view depending on the surgical stage. The camera view may be tracking the instruments or pointing at a particular anatomical structure. In our previous work [15], we propose a cognitive architecture based on a long-term memory that stores the robot's knowledge to provide the best camera view for each stage of the task. This work also included an episodic memory that takes into account particular preferences of different users. This work was improved in [16] with a navigation strategy that merges the advantages of a reactive instrument tracking with a proactive control based on a predefined camera behavior for each task stage. A reinforcement learning algorithm was used to learn the weight of each kind of control to the global behavior of the camera. This work revealed that this autonomous navigation of the camera improved the surgeon performance and did not require the interaction of the surgeon. However, this strategy requires an exhaustive hand-crafted model of the control strategy.

To solve this issue, learning from demonstration (LfD) arises as a natural means to transfer human knowledge to a robot that avoids the conventional manual programming. Essentially, a robot observes how a human performs a task (i.e., the demonstration) and then it autonomously reproduces the human behavior to complete the same task (i.e., the imitation). This approach is used in a wide variety of applications like rehabilitation and assistive robots [17], motion planning [18], intelligent autopilot systems [19], learning and reproduction gestures [20], or haptic guidance assistance for shared control applications [21,22]. In the field of surgical robotics, LfD has been used for different purposes. Reiley et al. [23] proposed the use of this method to train and reproduce robot trajectories from previous expert demonstrations, which were obtained using the Da Vinci surgical system. These trajectories were used to evaluate different surgeon skill levels (expert, intermediate and novice). On the other hand, van den Berg et al. [24] proposed the use of LfD to allow surgical robotic assistants to execute specific tasks with superhuman performance in terms of speed and smoothness. Using this approach, the Berkeley Surgical Robot was trained to tie a knot in a thread around a ring following a three-stage procedure. The results of this experiment demonstrated that the robot was able to successfully execute this task up to 7x faster than the demonstration. Recently, Chen et al. [25] propose the use of LfD combined with reinforcement learning methods to learn the inverse kinematics of a flexible manipulator from human demonstrations. Two surgical tasks were carried out to demonstrate the effectiveness of the proposed method.

This paper explores the use of LfD to guide the camera during laparoscopic surgery. A new approach to transfer human know-how from previous demonstrations is defined. It uses Gaussian mixture models (GMM) to generate a model of the task, which is later used to generate the camera motions by means of Gaussian mixture regression (GMR). This approach has been experimentally validated through a surgical robotic platform that is composed of three manipulators, which holds two instruments and the endoscopic camera. This platform has been used to train different behaviors of the camera during a peg transferring task, which is commonly used to train surgeons skill. The information provided by the training was used to create a GMM for this task. Later on, the same robotic platform was used to reproduce the task and provide the autonomous camera guidance from the GMM. Training and reproduction were evaluated in order to validate the proposed approach to be applied in surgical robotics.

This paper is organized as follows. Section 2 describes the autonomous camera guidance method based on a LfD approach. The experiments performed to validate the proposal are detailed in Section 3, and Section 4 presents the discussion and the future work.

2. Autonomous Camera Guidance

Figure 1 shows the general scenario of an abdominal laparoscopic surgery, with two surgical instruments and the camera, which tip positions are defined as $\vec{p_1} = (p_{1x}, p_{1y}, p_{1z})$, $\vec{p_2} = (p_{2x}, p_{2y}, p_{2z})$, and $\vec{c} = (c_x, c_y, c_z)$, respectively. The idea of the guidance approach is to teach the system how the camera moves depending on the surgical instrument positions at a given time, as they are the main reference to establish the viewpoint. Thus, we have to set a common reference frame to relate the instruments and the camera measures, which is represented as $\{O\}$ in the figure. The most natural choice is to set a global frame associated with an important location within the particular task for which the system is going to be trained, e.g., if we are training the system to move the camera in a cholecystectomy, then a natural reference would be the gallbladder, but if the task is a kidney transplantation then it is more reasonable to take the kidney as the reference. Hence, the global frame $\{O\}$ was chosen for each particular task, and it was set in an initial calibration process at the beginning of each set of demonstrations and reproductions of the task.

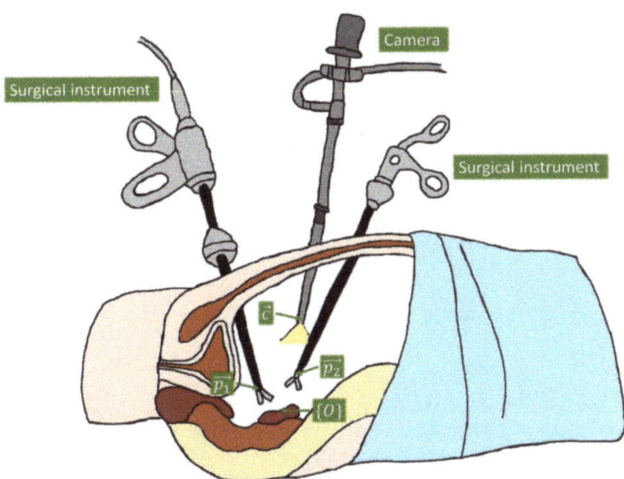

Figure 1. General scenario of an abdominal laparoscopic surgery.

The autonomous camera guidance method proposed in this paper followed the LfD approach shown in Figure 2, which was based on two stages: the first one, the off-line stage, created the robot knowledge base from human expert demonstrations; and the second one, the on-line stage, used that knowledge to generate the camera motion. During the off-line stage, an expert surgeon performed a set of demonstrations of the camera motion for a particular task, and the system stored the camera position and its relation to the current position of the surgical instruments, i.e., the camera position $\vec{c}$ for each tuple of instrument positions $(\vec{p_1}, \vec{p_2})$. The demonstrations can be carried out with the surgical robotic platform used in the experiments, or using other devices, such as joysticks. Moreover, conventional surgical tools could also be employed, using a tracking position sensor as it was done in [26] for surgical manoeuvre recognition purposes. The objective of this process, called know-how transferring, was to create a knowledge base that stores the behavioral model of the camera (ρ). Then, during the on-line stage, the motion generation module took the previously trained model ρ and the current position of the surgical instruments, $\vec{p_1}$ and $\vec{p_2}$, to update the camera location, $\bar{c}$. To ensure the safety of the patient in a real surgery, the system must include a human–machine interface (HMI) that allows the surgeon to take control of the camera in case of an undesirable motion, and a supervisor system that guarantees that the camera moves inside a safety region.

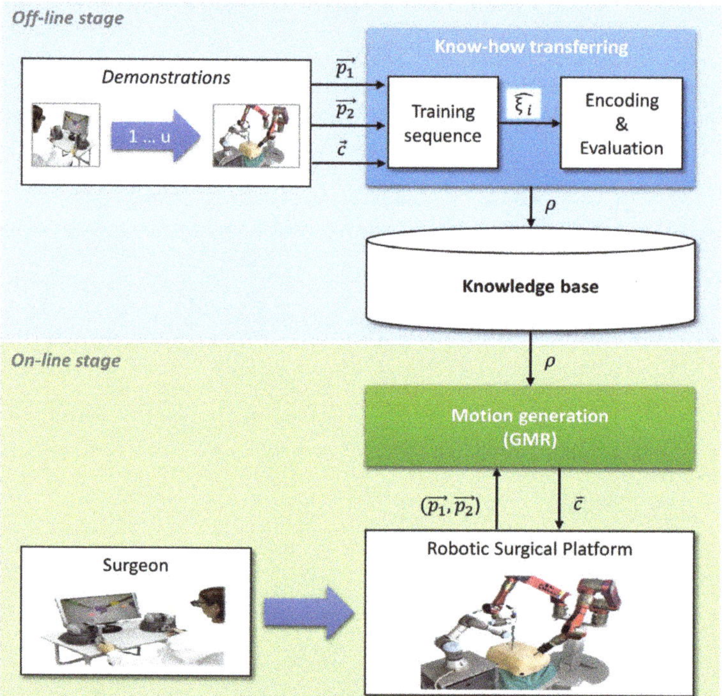

Figure 2. Learning for demonstration approach for autonomous camera guidance.

2.1. Know-How Transferring

Know-how transferring is the process through which an expert surgeon demonstrates the camera motion for a particular surgical task. Starting from a data tuple that contains the position of the instruments and camera, which can be defined as follows,

$$\xi = (p_{1x}, p_{1y}, p_{1z}, p_{2x}, p_{2y}, p_{2z}, c_x, c_y, c_z) \tag{1}$$

each demonstration i generated a training sequence of k data tuples as:

$$\hat{\xi}_i(k) = \xi(1), ..., \xi(k), 1 \leq i \leq u. \tag{2}$$

Once the demonstrations were performed, the generated training sequences were encoded into a GMM ρ. A GMM is a probabilistic model that is able to include the training sequences in a set of N Gaussians, whose mixture covers the training sequences. So, a GMM can be defined as:

$$\rho = \{\pi_n, \mu_n, \Sigma_n\}_{n=1}^{N}, \tag{3}$$

where a Gaussian n is represented by π_n, which is the weight of each n Gaussian on the GMM, and μ_n and Σ_n are the mean and covariance matrix of the Gaussian, respectively.

The training of a GMM was carried out by means of the expectation-maximization (EM) algorithm [27]. The EM is an iterative algorithm that estimates the values of the N Gaussians of the training sequences, maximizing the likelihood of the training sequences belonging to the encoded GMM. The EM algorithm can be defined as follows, where the inputs are all the u training sequences and the number of Gaussians N:

$$\rho = \text{EM}(\hat{\xi}_{1...u}, N). \tag{4}$$

Once the model has been trained, it can be evaluated through the Bayesian information criterion (BIC) as follows:

$$\text{BIC} = -L((k)) + \frac{n_p}{2}\log(k) \tag{5}$$

where L is defined as:

$$L(\hat{\xi}(k)) = \sum_{j=1}^{k} \log\left(P\left(\xi\left(j\right)\right)\right) \tag{6}$$

with:

$$P(\xi) = \sum_{n=1}^{N} P(n)P(\xi|n) \tag{7}$$

In this equations, $P(n)$ is the prior probability and $P(\xi|n)$ is the conditional probability density function, both defined as:

$$P(n) = \pi_n, \tag{8}$$
$$P(\xi|n) = N(\xi, \mu_n, \Sigma_n). \tag{9}$$

Finally, in Equation (5), n_p is a variable used to penalize the score taking into consideration the dimension of the tuple, D, and the number of Gaussians, N, as follows:

$$n_p = (N-1) + N(D + \tfrac{1}{2}D(D+1)). \tag{10}$$

This method provided a score that was used to choose the best number of Gaussians N. The lower the score, the better the model fitness was [28].

2.2. Motion Generation

Once the model has been trained within ρ (Equation (3)) and evaluated, the camera motion could be extracted through GMR [27]. For this purpose, the parameters of ρ can be represented as:

$$\mu_n = \begin{bmatrix} \mu_n^p \\ \mu_n^c \end{bmatrix}, \quad \Sigma_n = \begin{bmatrix} \Sigma_n^p & \Sigma_n^{pc} \\ \Sigma_n^{cp} & \Sigma_n^c \end{bmatrix}. \tag{11}$$

These parameters were used to obtain the camera position using GMR as stated in Equation (12), where $\bar{c} = (c_x, c_y, c_z)$ is the camera position generated for a particular position of the instruments $\vec{p_1} = (p_{1x}, p_{1y}, p_{1z})$ and $\vec{p_2} = (p_{2x}, p_{2y}, p_{2z})$:

$$\bar{c} = \sum_{n=1}^{N} P(n|(p_{1x}, p_{1y}, p_{1z}, p_{2x}, p_{2y}, p_{2z}) \left[\mu_n^c + \frac{\Sigma_n^{pc}}{\Sigma_n^p}((p_{1x}, p_{1y}, p_{1z}, p_{2x}, p_{2y}, p_{2z}) + \mu_n^p) \right], \qquad (12)$$

where

$$P(n|(p_{1x}, p_{1y}, p_{1z}, p_{2x}, p_{2y}, p_{2z})) = \frac{P(n)P((p_{1x}, p_{1y}, p_{1z}, p_{2x}, p_{2y}, p_{2z})|n)}{\sum_{j=1}^{N} P(j)P((p_{1x}, p_{1y}, p_{1z}, p_{2x}, p_{2y}, p_{2z})|j)}. \qquad (13)$$

These equations can be used every sample time to generate the camera position $\bar{c}$ and send it to the surgical robotic platform in order to autonomously move the camera to the corresponding position.

3. Experimental Results

The autonomous camera guidance method, proposed in the previous section, has been validated through two experiments, each one for each stage of training and task reproduction. The training stage showed how a surgical task could be trained using GMM, and the second one, task reproduction, showed the camera behavior based on the previously learned knowledge to perform autonomous camera motions. These experiments have been carried out using the surgical robotic platform described below.

3.1. Experimental Scenario

Figure 3 shows the experimental scenario used to perform the validation of this work. The surgical robotic platform was composed of three arms; two of them that make up the CISOBOT platform, in charge of the teleoperation of the surgical instruments, and a UR3 robot, from Universal Robots, with a commercial 2D endoscope attached at its end-effector. The CISOBOT is an experimental platform developed at the University of Malaga, which is composed of two customized six degrees-of-freedom manipulators with a laparoscopic grasper tool attached at their end-effectors [29]. These robots are teleoperated using the master console that is shown in Figure 3, which is composed of two haptic devices and a monitor that displays the image of the camera. The surgeon used two commercial haptic devices, without force feedback, to teleoperate the robots (Phantom Omni, Sensable Technologies), and during the training stage, an additional haptic was used to move the camera. The master console was placed in the same room as the robotic platform, but the surgeon did not have direct vision of the task, so the only visual feedback he/she had is the image of the camera.

The instruments were inserted in an abdomen simulator with the experimental board shown in Figure 4 inside. It was a commercial pick-and-place pegboard developed by Medical Simulator used to train basic laparoscopic skills. This pegboard was 125 mm × 125 mm, and it had six cylindrical rubber rings, which the user (usually a novice surgeon) had to transfer from one peg to another. This was one of the five tasks described in the SAGES manual of skills in laparoscopic surgery [30] to measure technical skills during basic laparoscopic surgical maneuvers. The main purpose of this task was to exercise depth perception in a 2D environment, where the only visual feedback we had was the image of the camera. Thus, this was a suitable task to evaluate the autonomous camera guidance method proposed in this work, as the camera view was crucial for the successful performance of the task.

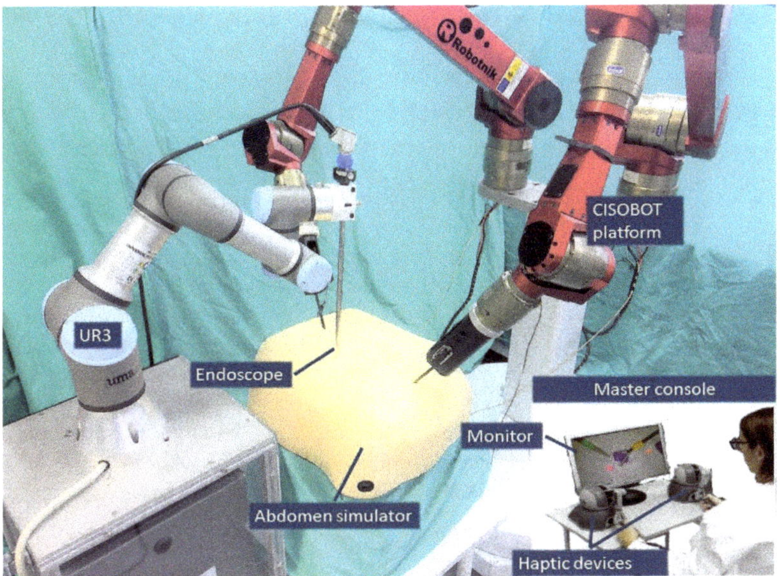

Figure 3. Experimental surgical robotic platform. It was composed of a teleoperation master console, an endoscope that was handled by a UR3 robot, and the CISOBOT platform.

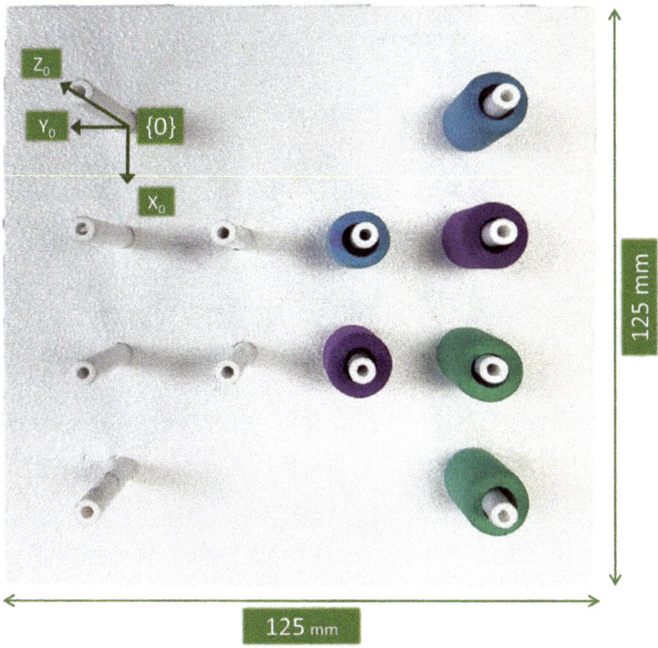

Figure 4. Experimental board. It had 12 pegs and the objective was to move cylindrical rubber rings from one peg to another.

For this task, the global frame $\{O\}$ was set at the top left peg of the pegboard, as shown in Figure 4. The global frame calibration process comprised of two steps: first, the user touched the top left peg with the tip of the right surgical tool to establish the relation between the pegboard and the CISOBOT; and second, the user touched the tip of the camera with the same tool to relate the CISOBOT and the UR3 robot. Afterward, these relations were stored in the system, and all the data measured during the performance of the experiments was transformed to the global frame $\{O\}$.

3.2. Data Acquisition

The implementation of the method for autonomous camera guidance in laparoscopic surgery follows the software/hardware architecture shown in Figure 5. It was based on the open-source framework ROS [31], which allows easy communication between the different components of the system. The two main nodes of the software architecture are the *tool teleoperation node* and the *UR3 robot node*, both running at 125 Hz. The CISOBOT platform was controlled by real-time hardware (NI-PXI, http://www.ni.com/en-nz/shop/pxi.html) that provided natural teleoperation of the surgical tools. This control was integrated into a ROS node that published the position of the tools to the rest of the system. On the other hand, the camera had two ways of operating; during the off-line stage, the camera was teleoperated using ROS to communicate the Phantom Omni with the UR3 robot; and during the on-line stage, the camera followed an autonomous guidance according to the motion generated by the GMR model, which was implemented in a MATLAB environment.

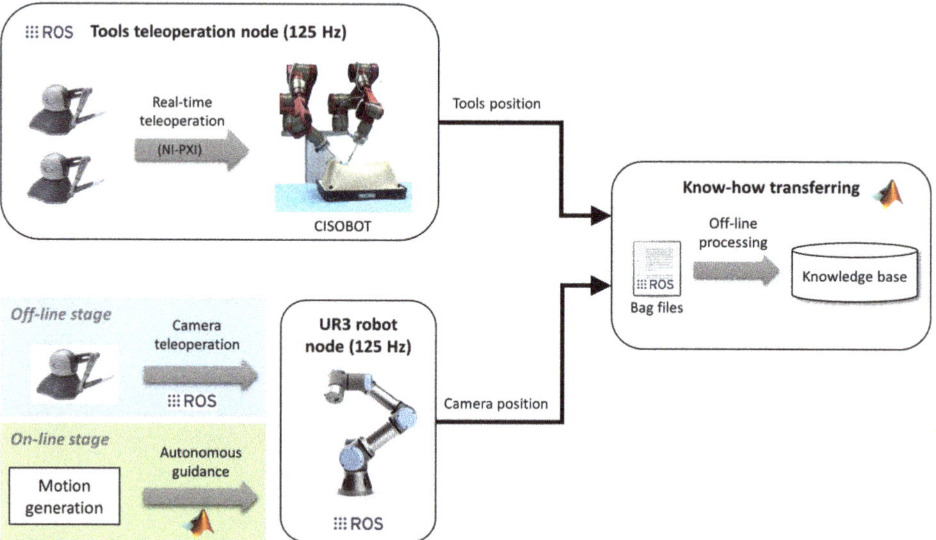

Figure 5. SW/HW architecture for the autonomous camera guidance method.

Finally, during the know-how transferral process, the training data was stored in a bag file format, which allowed recording synchronized data published by different ROS nodes for off-line analysis. Hence, the bag files recorded the surgical tool positions from the CISOBOT node and the camera position from the UR3 node. Then, that data was encoded in MATLAB to generate the knowledge base containing the GMM of the camera behavior.

The training and reproduction data, as well as the source code, implemented to perform this work have been published as an open repository in github (https://github.com/spaceuma/LfDCameraGuidance).

3.3. Training

The training stage of the autonomous camera guidance approach consisted of teaching the system how to move the camera during the pick-and-place task described above. Hence, the training needed two people: the main surgeon that moved the tools and a surgeon assistant that handled the camera. The task consisted of picking the rings with one of the tools, transferring it to the other tool, and placing it in another peg of the pegboard. As the objective was to learn a general behavior of the camera, there was no predefined order to pick-and-place the rings. The task began with the six rings placed in the positions shown in Figure 4. Then the surgeon can freely choose both which ring to pick up and the peg to place them. When a ring falls, the surgeon can pick it again with one of the tools, unless it falls out of the workspace of the robots. When finished, the rings were placed back to the original position by hand, and the task was repeated again. In total, the training dataset contained k = 2616.34 s · 125 Hz = 327.043 tuples ζ, which corresponded to the duration of the training multiplied by the recording rate.

Before starting the training of the system, the surgeon and the assistant agreed on the behavior of the camera depending on the instruments relative position. This behavior had been defined so that the method was as general as possible, trying to minimize the effect of the particular preferences of different surgeons. Therefore, after this a priori conversation, the surgeon was asked to not provide additional on-line instructions to the assistant to modify the camera view. This way, the behavior of the camera followed the general guidelines independently of the surgeon's preferences. The agreed qualitative behavioral model of the camera can be divided into the following guidelines:

- Instruments tracking: in the horizontal plane (xy plane), the camera must follow the tools trying to keep always both instruments in the field of view but focusing on the active tool in case only one of them is moving. In the vertical plane (z plane), the camera should follow the tools from a certain distance that provides a suitable trade-off between field of view and zoom. Particular grades of zoom are performed by following the inward trajectories of the instruments.
- Zoom-out: in a typical laparoscopic procedure, surgeons operate in very specific areas and they need the camera to focus in that particular zone. However, sometimes they need to have a global vision of the operating area, i.e., to zoom-out the camera from the surgical tools. As this fact depends on the needs of the surgeon at a particular time, the surgeon and the assistant need a non-verbal signal to teach the system when to zoom-out the camera. For this task, the system has been trained to zoom-out when both tools make a synchronized outwards motion.

Figure 6 shows an example of the training trajectories and the GMM fitness for the xy plane with respect to the reference frame $\{O\}$ for both instruments. The trajectories are represented with blue lines and the Gaussians within the GMM are represented with green ellipses. Finally, the pegboard area is represented by a grey square. As shown, the Gaussians covered the trained trajectories correctly. Comparing Figure 6a,b, it can be appreciated how the left tool covered a greater part of the left area of the pegboard, while the right reached the zone further to the right. Trajectories out of the pegboard correspond with failure situations, in which one ring has fallen out of the tools outside the pegboard limits.

Similarly, Figure 6c shows the trained camera trajectories and the learned GMM model. In this figure, the fulcrum point of the camera is represented by a red circle. This point has been chosen so that the camera view covers all the pegboard from its outer position. As the camera does not enter into the abdomen as much as the surgical tools, it does not reach as far as x and y positions as the instruments.

The number of Gaussians N is an important parameter within a GMM, so its appropriate choice is critical for the system training. Indeed, a low N would provide a poorly trained model and a high N would increase the CPU time, which would affect the real-time constraints. To decide the best number of Gaussians, the BIC score was defined in Equation (5). Figure 7 represents the BIC score for $1 \leq N \leq 30$. As shown, the score decreases as N increases, which means that the model better fits the training sequence as the number of Gaussians is increased. Therefore, the main limitation was the CPU processing time for real-time constraints. Using the proposed architecture (Figure 5), which used MATLAB Simulink Desktop Real-Time and ROS, a value $N = 20$ was reached. Higher values than this one made it impossible to be executed on the platform.

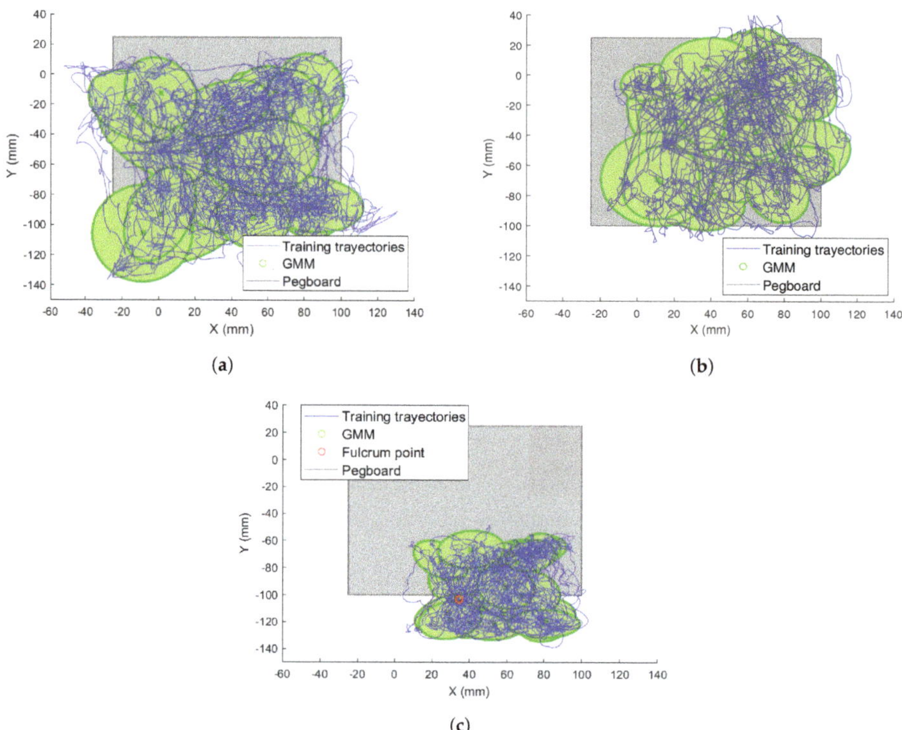

(a)   (b)

(c)

Figure 6. Trajectories and Gaussian mixture models (GMM) model in the xy plane during the training stage: (**a**) left tool; (**b**) right tool; and (**c**) camera.

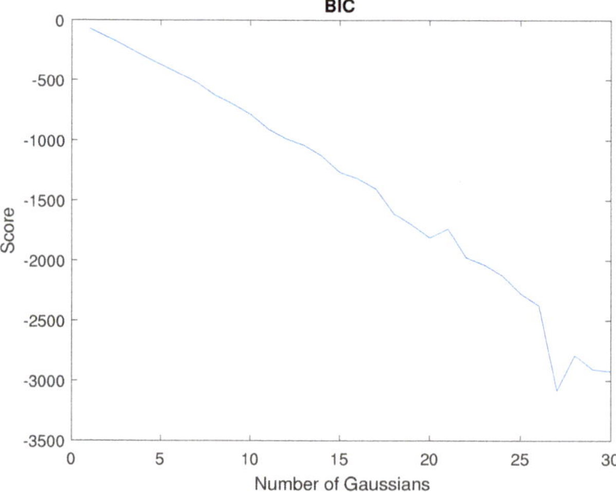

Figure 7. Bayesian information criterion (BIC) score for the generated GMM using the training sequence. The score is decreased (improved) as the number of gaussians are increased.

3.4. Task Reproduction

To validate the autonomous camera guidance method, the main surgeon has repeated the pick-and-place task described above during a period of 10 min, recording a total of 78.712 data tuples. The reproduction of the task began with the rings placed in the same positions as at the beginning of the training, i.e., as shown in Figure 4. Then, during the task reproduction, the surgeon can freely choose which ring to pick and the location to place it. This way, it is assured that the training and the reproduction tasks were not performed in the same order. During this experiment, the camera robot followed autonomously the previously trained behavior. Figure 8 shows the tools and the camera trajectories during this validation experiment. As it can be seen, the tools moved within the pegboard area, and the camera moved according to the trained positions of Figure 6c. During the whole task reproduction, the system provided a camera view that allowed the surgeon to perform the task without the necessity to manually change the camera position. A video showing a representational part of this task reproduction can be found in the Supplementary Materials.

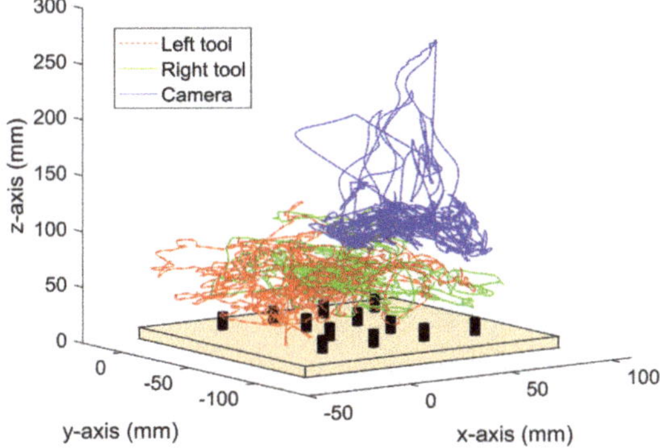

Figure 8. Tools and camera trajectories during the pick-and-place task reproduction.

Figure 9 shows a comparison of the instrument tracking behavior during the training stage of the system and the task reproduction. The results are divided in the trajectories in the horizontal xy plane (Figure 9a) and in the vertical z plane (Figure 9b). In the horizontal plane, it can be seen that the camera moved within the trained area. This figure also shows that during the reproduction, the left tool reached an area on the lower-left of the pegboard that was not trained during the off-line stage. This is due to a failure during a ring transfer in which the ring fells out of the pegboard. As a consequence, the camera moved to an area further in the left than the trained behavior, which allowed it to keep the instruments within the camera view.

Regarding the vertical tracking, Figure 9b shows that there were two behaviors that are clearly identifiable: the zoom-out gesture described in the previous section, in which the camera got to the highest z positions, and the tracking in the z-axis during the normal performance of the task. Figure 10 shows examples of the zoom-out gesture during the training (Figure 10a) and the reproduction (Figure 10b) of the task. In both figures, a time window of the experiments are represented, and the particular instants of time in which the zoom-out gesture occurs are marked with shaded areas. At these instants, it can be appreciated how both tools raise in the z-axis, and the camera makes an outwards motion in response. Comparing the behavior during the training and the reproduction, it can be noted that during the training, the camera zooms out at a mean z position of 292 mm, while during the reproduction the camera gets a mean z value of 260 mm, around 30 mm less than the trained behavior. This has to do with the motion of the tools during both experiments. Analyzing the data, during the training stage, the tools reached a higher position than during the reproduction, which is the reason why the camera had a lower position in the latter case. However, Figure 10 shows how the zoom-out gesture is correctly detected by the system, and the trained behavior is performed.

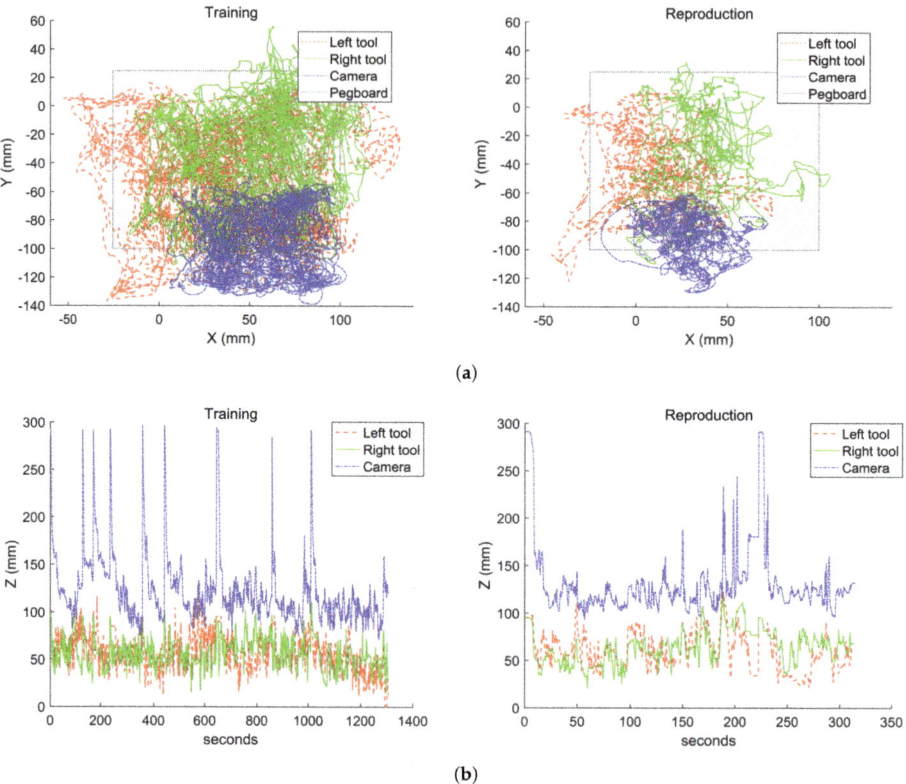

Figure 9. Comparison between the training and the reproduction experiments for the instruments tracking behavior: (**a**) tracking of the instruments in the horizontal plane, and (**b**) tracking in the vertical plane.

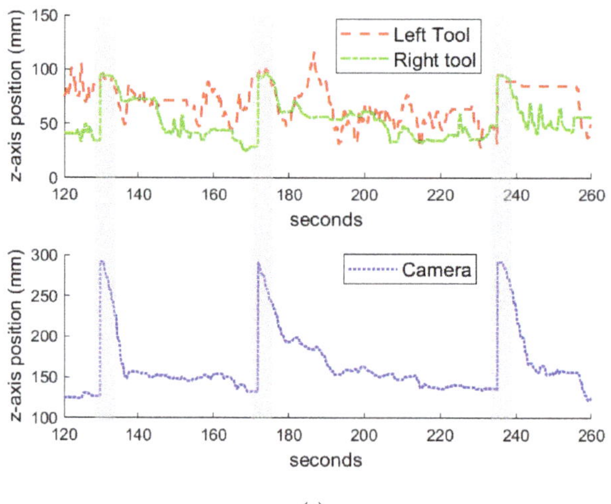

(**a**)

Figure 10. *Cont.*

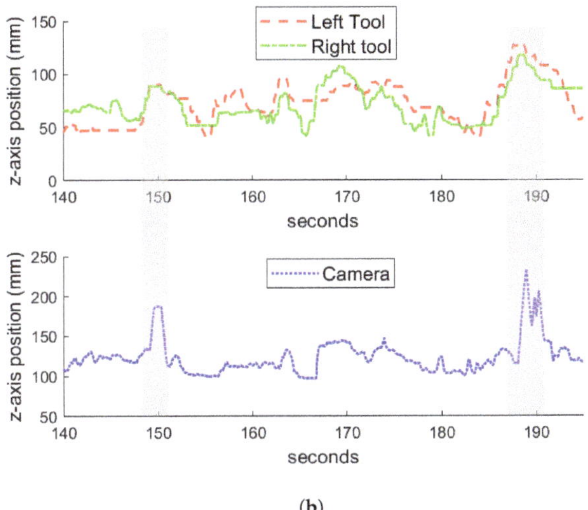

(**b**)

Figure 10. Zoom-out gesture during the (**a**) training and the (**b**) reproduction experiments. The shadow areas represent the moment in time in which the gesture occurs.

4. Discussion

This paper shows the feasibility of using learning from demonstration for autonomous camera guidance in laparoscopic surgery. Peg transferring was selected as a suitable task that demonstrated the surgeon skills. Around 43 min were spent to generate enough information to train a GMM with 20 Gaussians, which was considered enough to perform camera guidance with a 125 Hz real-time system. During the training stage, two behaviors were considered: instrument tracking and zoom-out. They were evaluated by repeating the same task for a period of 10 min. Both the attached video and figures, show that the previously defined behaviors were accomplished during the task reproduction stage, demonstrating that camera guidance is a suitable task to be carried out autonomously. However, there are several issues that remain to be investigated. The first one is related to the global frame pose for the task. In the experimental scenario, a well known global reference was used, i.e., the top left cylindrical rubber ring. However, in real surgery, the global reference frame would be difficult to choose, as it must be something fixed inside the abdominal cavity, and also, the position and dimension of the organs are different depending on the patient. The second issue is related to the task recognition. As stated in Section 1, there is a lot of work related to the detection of the surgical procedure stage in order to fix some parameters. It would be desirable to use these methods to detect the task that is being carried out by the surgeon, and therefore, apply the correct camera guidance model. Finally, the autonomous camera guidance method proposed in this work should be tested in a more complex task performed by expert surgeons. Moreover, in order to include it in a real surgery, the system must include a supervisor system that guarantees the patient's safety. The solution to these issues is proposed as future works.

Supplementary Materials: The following are available online at http://www.mdpi.com/2079-9292/8/2/224/s1, Video S1: Representational part of the task reproduction.

Author Contributions: Conceptualization, I.R.-B. and C.J.P.-d.-P.; methodology, C.J.P.-d.-P.; software, C.J.P.-d.-P., E.B.; validation, I.R.-B., C.L.-C.; writing—original draft, I.R.B., C.J.P.-d.-P.; funding acquisition, V.F.M.

Funding: This research was funded by the Spanish Ministry of Economy and Competitiveness under the grant number DPI2016-80391-C3-R.

Conflicts of Interest: The authors declare no conflict of interest.

Abbreviations

The following abbreviations are used in this manuscript:

LfD	Learning from demonstration
GMM	Gaussian mixture model
EM	Expectation-maximization
BIC	Bayesian information criterion
GMR	Gaussian mixture regression
HMI	Human–machine interface
SW/HW	Software/Hardware

References

1. da Vinci Surgery—Minimally Invasive Robotic Surgery with the da Vinci Surgical System. Available online: https://www.davincisurgery.com/da-vinci-surgery/da-vinci-surgical-system/ (accessed on 17 December 2018).
2. Omote, K.; Feussner, H.; Ungeheuer, A.; Arbter, K.; Wei, G.Q.; Siewert, J.; Hirzinger, G. Self-guided robotic camera control for laparoscopic surgery compared with human camera control. *Am. J. Surg.* **1999**, *177*, 321–324. [CrossRef]
3. Muñoz, V.; Gómez-de Gabriel, J.; Garcia-Morales, I.; Fernández-Lozano, J.; Morales, J. Pivoting motion control for a laparoscopic assistant robot and human clinical trials. *Adv. Robot.* **2005**, *19*, 694–712. [CrossRef]
4. Stolzenburg, J.U.; Franz, T.; Kallidonis, P.; Minh, D.; Dietel, A.; Hicks, J.; Nicolaus, M.; Al-Aown, A.; Liatsikos, E. Comparison of the FreeHand® robotic camera holder with human assistants during endoscopic extraperitoneal radical prostatectomy. *BJU Int.* **2011**, *107*, 970–974. [CrossRef] [PubMed]
5. Noonan, D.; Mylonas, G.; Shang, J.; Payne, C.; Darzi, A.; Yang, G.Z. Gaze contingent control for an articulated mechatronic laparoscope. In Proceedings of the 2010 3rd IEEE RAS & EMBS International Conference on Biomedical Robotics and Biomechatronics, Tokyo, Japan, 26–29 September 2010; pp. 759–764. [CrossRef]
6. Casals, A.; Amat, J.; Laporte, E. Automatic Guidance of an Assistant Robot in Laparoscopic Surgery. In Proceedings of the IEEE International Conference on Robotics and Automation Minneapolis, Minneapolis, MN, USA, 22–28 April 1996; pp. 895–900.
7. Ali, S.; Reisner, L.; King, B.; Cao, A.; Auner, G.; Klein, M.; Pandya, A. Eye gaze tracking for endoscopic camera positioning: An application of a hardware/software interface developed to automate Aesop. *Stud. Health Technol. Inform.* **2008**, *132*, 4–7. [PubMed]
8. Azizian, M.; Khoshnam, M.; Najmaei, N.; Patel, R. Visual servoing in medical robotics: A survey. Part I: Endoscopic and direct vision imaging—Techniques and applications. *Int. J. Med. Robot. Comput. Assist. Surg.* **2014**, *10*, 263–274. [CrossRef] [PubMed]
9. Sarikaya, D.; Corso, J.J.; Guru, K.A. Detection and Localization of Robotic Tools in Robot-Assisted Surgery Videos Using Deep Neural Networks for Region Proposal and Detection. *IEEE Trans. Med. Imaging* **2017**, *36*, 1542–1549. [CrossRef] [PubMed]
10. Probst, T.; Maninis, K.K.; Chhatkuli, A.; Ourak, M.; Poorten, E.V.; Van Gool, L. Automatic Tool Landmark Detection for Stereo Vision in Robot-Assisted Retinal Surgery. *IEEE Robot. Autom. Lett.* **2018**, *3*, 612–619. [CrossRef]
11. Wang, S.; Raju, A.; Huang, J. Deep learning based multi-label classification for surgical tool presence detection in laparoscopic videos. In Proceedings of the International Symposium on Biomedical Imaging, Melbourne, VIC, Australia, 18–21 April 2017; pp. 620–623. [CrossRef]
12. Weede, O.; Monnich, H.; Muller, B.; Worn, H. An intelligent and autonomous endoscopic guidance system for minimally invasive surgery. In Proceedings of the 2011 IEEE International Conference on Robotics and Automation, Shanghai, China, 9–13 May 2011; pp. 5762–5768. [CrossRef]

13. King, B.; Reisner, L.; Pandya, A.; Composto, A.; Ellis, R.; Klein, M. Towards an Autonomous Robot for Camera Control During Laparoscopic Surgery. *J. Laparoendosc. Adv. Surg. Tech.* **2013**, *23*, 1027–1030. [CrossRef] [PubMed]

14. Ko, S.; Kim, J.; Kwon, D.S.; Lee, W.J. Intelligent interaction between surgeon and laparoscopic assistant robot system. In Proceedings of the ROMAN 2005, IEEE International Workshop on Robot and Human Interactive Communication, Nashville, TN, USA, 13–15 August 2005; pp. 60–65. [CrossRef]

15. Rivas-Blanco, I.; Estebanez, B.; Cuevas Rodríguez, M.; Muñoz, V.; Bauzano, E.; Munoz, V. Towards a cognitive camera robotic assistant. In Proceedings of the 5th IEEE RAS & EMBS International Conference on Biomedical Robotics and Biomechatronics, Sao Paulo, Brazil, 12–15 August 2014; pp. 739–744. [CrossRef]

16. Rivas-Blanco, I.; Lopez-Casado, C.; Perez-del Pulgar, C.J.; Garcia-Vacas, F.; Fraile, J.C.; Munoz, V.F. Smart Cable-Driven Camera Robotic Assistant. *IEEE Trans. Hum. Mach. Syst.* **2018**, *48*, 183–196. [CrossRef]

17. Lauretti, C.; Cordella, F.; Guglielmelli, E.; Zollo, L. Learning by Demonstration for Planning Activities of Daily Living in Rehabilitation and Assistive Robotics. *IEEE Robot. Autom. Lett.* **2017**, *2*, 1375–1382. [CrossRef]

18. Wang, H.; Chen, J.; Lau, H.Y.K.; Ren, H. Motion Planning Based on Learning From Demonstration for Multiple-Segment Flexible Soft Robots Actuated by Electroactive Polymers. *IEEE Robot. Autom. Lett.* **2016**, *1*, 391–398. [CrossRef]

19. Baomar, H.; Bentley, P.J. An Intelligent Autopilot System that learns piloting skills from human pilots by imitation. In Proceedings of the 2016 International Conference on Unmanned Aircraft Systems (ICUAS), Arlington, VA, USA, 7–10 June 2016; pp. 1023–1031. [CrossRef]

20. Calinon, S.; D'halluin, F.; Sauser, E.; Caldwell, D.; Billard, A. Learning and Reproduction of Gestures by Imitation. *IEEE Robot. Autom. Mag.* **2010**, *17*, 44–54. [CrossRef]

21. Perez-del Pulgar, C.J.; Smisek, J.; Munoz, V.F.; Schiele, A. Using learning from demonstration to generate real-time guidance for haptic shared control. In Proceedings of the 2016 IEEE International Conference on Systems, Man, and Cybernetics (SMC), Budapest, Hungary, 9–12 Ocotber 2016; pp. 003205–003210. [CrossRef]

22. Power, M.; Rafii-Tari, H.; Bergeles, C.; Vitiello, V.; Yang, G.-Z. A cooperative control framework for haptic guidance of bimanual surgical tasks based on Learning From Demonstration. In Proceedings of the 2015 IEEE International Conference on Robotics and Automation (ICRA), Seattle, WA, USA, 26–30 May 2015; pp. 5330–5337. [CrossRef]

23. Reiley, C.; Plaku, E.; Hager, G. Motion generation of robotic surgical tasks: Learning from expert demonstrations. In Proceedings of the 2010 Annual International Conference of the IEEE Engineering in Medicine and Biology, Buenos Aires, Argentina, 31 August–4 September 2010; pp. 967–970. [CrossRef]

24. Van den Berg, J.; Miller, S.; Duckworth, D.; Hu, H.; Wan, A.; Xiao-Yu, F.; Goldberg, K.; Abbeel, P. Superhuman performance of surgical tasks by robots using iterative learning from human-guided demonstrations. In Proceedings of the 2010 IEEE International Conference on Robotics and Automation, Anchorage, AK, USA, 3–7 May 2010; pp. 2074–2081. [CrossRef]

25. Chen, J.; Lau, H.; Xu, W.; Ren, H. Towards transferring skills to flexible surgical robots with programming by demonstration and reinforcement learning. In Proceedings of the 2016 Eighth International Conference on Advanced Computational Intelligence (ICACI), Chiang Mai, Thailand, 14–16 February 2016; pp. 378–384. [CrossRef]

26. Estebanez, B.; Jimenez, G.; Muñoz, V.; Garcia-Morales, I.; Bauzano, E.; Molina, J. Minimally invasive surgery maneuver recognition based on surgeon model. In Proceedings of the 2009 IEEE/RSJ International Conference on Intelligent Robots and Systems, St. Louis, MO, USA, 10–15 October 2009; pp. 5522–5527. [CrossRef]

27. Calinon, S. *Robot Programming by Demonstration*; EPFL Press: Lausanne, Switzerland, 2009.

28. Chen, S.S.; Gopalakrishnan, P.S. Clustering via the Bayesian information criterion with applications in speech recognition. In Proceedings of the 1998 IEEE International Conference on Acoustics, Speech and Signal Processing, Seattle, WA, USA, 15 May 1998; Volume 2, pp. 645–648. [CrossRef]

29. Del Pulgar, C.P.; Garcia-Morales, I.; Blanco, I.R.; Munoz, V.F. Navigation Method for Teleoperated Single-Port Access Surgery With Soft Tissue Interaction Detection. *IEEE Syst. J.* **2016**, *12*, 1381–1392. [CrossRef]

30. Choy, I.; Okrainec, A. Fundamentals of Laparoscopic Surgery-FLS. In *The SAGES Manual of Quality, Outcomes and Patient Safety*; Springer: Boston, MA, USA, 2012; pp. 461–471. [CrossRef]
31. Available online: https://www.ros.org (accessed on 17 December 2018).

Article

On Inferring Intentions in Shared Tasks for Industrial Collaborative Robots

Alberto Olivares-Alarcos *, Sergi Foix and Guillem Alenyà

Institut de Robòtica i Informàtica Industrial, CSIC-UPC, Llorens i Artigas 4-6, 08028 Barcelona, Spain; sfoix@iri.upc.edu (S.F.); galenya@iri.upc.edu (G.A.)
* Correspondence: aolivares@iri.upc.edu; Tel.: +34-93-4010934

Received: 29 September 2019; Accepted: 1 November 2019; Published: 7 November 2019

Abstract: Inferring human operators' actions in shared collaborative tasks plays a crucial role in enhancing the cognitive capabilities of industrial robots. In all these incipient collaborative robotic applications, humans and robots not only should share space, but also forces and the execution of a task. In this article, we present a robotic system that is able to identify different human's intentions and to adapt its behavior consequently, only employing force data. In order to accomplish this aim, three major contributions are presented: (a) a force based operator's intention recognition system based on data from only two users; (b) a force based dataset of physical human–robot interaction; and (c) validation of the whole system with 15 people in a scenario inspired by a realistic industrial application. This work is an important step towards a more natural and user-friendly manner of physical human–robot interaction in scenarios where humans and robots collaborate in the accomplishment of a task.

Keywords: industrial collaborative robots; shared robotic tasks; physical human–robot interaction; human intention recognition; time series classification

1. Introduction

Currently, there is a rising trend towards smart factories where all the involved entities cooperate and communicate with each other. This is often referred to as Industry 4.0 or the fourth industrial revolution. Settling this aim for the industrial robotics sector would require freeing robots from their current work cells, closer to operators, compromising human safety [1,2]. In the interest of overcoming those safety issues, over the last few years, collaborative robots or cobots have emerged [3–5]. These robots are specifically designed for direct interaction with a human within a defined collaborative workspace [6]. Collaborative robots have meant great progress towards a safer coexistence of operators and industrial robots. Nevertheless, scenarios where humans and robots exchange forces and share the execution of a task require the use of robots equipped with complex cognitive capabilities [7]. Bauer et al. [8] proposed five levels of cooperation between robots and humans (see Figure 1). The authors stated that most of the current real applications of industrial robots are based on the cooperation levels coexistence and synchronized [9,10]. Driven by the lack of applications where more complex levels of cooperation are addressed, we propose a scenario based on the fifth level, collaboration. Figure 2 depicts the proposed setup, where a human and a robot exchange forces while sharing the execution of a task inspired by a realistic industrial scenario.

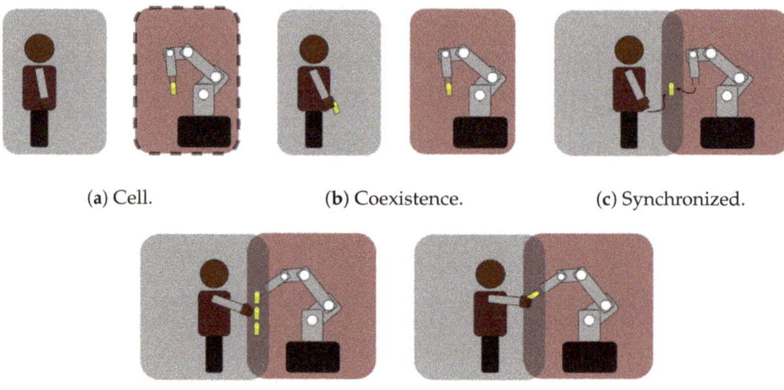

(**a**) Cell.   (**b**) Coexistence.   (**c**) Synchronized.

(**d**) Cooperation.   (**e**) Collaboration.

Figure 1. Human–robot cooperation levels in industrial environments. (**a**) The level cell involves no collaboration at all; the robot remains held inside a work cell. (**b**) Coexistence removes the cell, but humans and robots do not share the workspace yet. (**c**) Synchronized allows the sharing of the workspace, but never at the same time; humans and robots operate in a synchronized manner. (**d**) At the level cooperation, the task and the workspace are shared, but humans and robots do not physically interact. (**e**) The level collaboration considers full collaboration where operators and robots exchange forces.

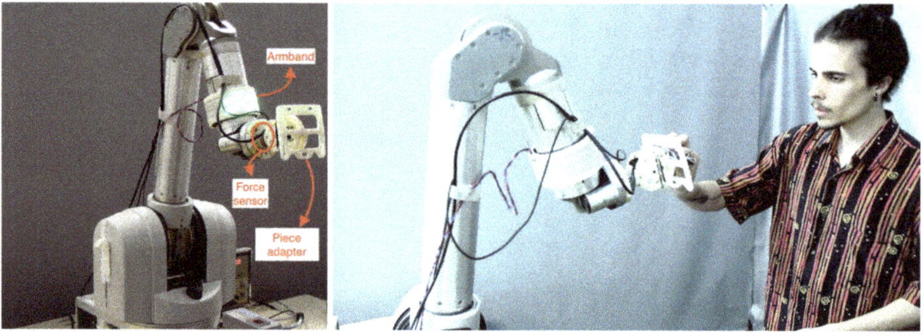

(**a**) Robot hardware components.   (**b**) User's pose with respect to the robot.

Figure 2. Proposed scenario inspired by an industrial collaborative robotic task in which the robot adapts its state to the human's intention. (**a**) The force sensor is used to infer the human's intention; the armband is used to inform the user about the robot's internal state; and the piece adapter eases the grasping of the object. (**b**) While the robot holds the object, the human performs a frontal polishing of it.

In a real industrial environment, operators tend to suffer from injuries related to the usual repetitive tasks involved in their daily duties. In our scenario, it is important to reduce as much as possible very mechanical movements and let the users interact with the robot through more natural kinds of gestures. Moreover, the cooperation between the human and the robot during repetitive physical human–robot interactions should be fluent [1,6]. Based on our experience, one second is the maximum amount of time for an efficiently responsive human–robot collaboration. In industrial surroundings, there is much heterogeneous contextual information that can have an effect on or modify the progress of a task. In future work, we would like to benefit from using that contextual information. Therefore, it would be desirable that our machine learning approach be able to cope with not only temporal sequences, but also other types of environmental variables. To sum up, natural interaction, fast prediction, and contextual variables will play a relevant role for the data gathering and the selection of the most appropriate approach.

The main contributions presented within this work are:

- Force based operator's intention inference. We implemented two different approaches, and both were thoroughly evaluated and compared. Finally, one of them was selected and used during the validation with users. Inference time and the possibility of including contextual information were considered for the comparison. The first approach consisted of a k-nearest neighbor classifier, which uses as the metric dynamic time warping. In this case, the time series data are directly fed to the classifier. The second approach was based on dimensionality reduction together with a support vector machine classifier. The reduction was performed over the concatenation of all force axes of the raw time series.
- Force based dataset of physical human–robot interaction. Due to the lack of similar existent datasets, we present a novel dataset containing force based information extracted from natural human–robot interactions. Geared towards the inference of operators' intentions, the dataset comprises labeled signals from a force sensor. We aimed to generalize from a few users to several. Therefore, our dataset was only recorded with two users. Indeed, this is compliant with industrial environments in which the system should be used by new operators, preferably with no need for retraining.
- Validation in a use-case inspired by a realistic industrial collaborative robotic scenario. The performance of the selected approach was evaluated in an experiment with fifteen users, who received a short explanation of the collaborative task to execute. The goal of the shared task was to inspect and polish a manufacturing piece where the robot adapted to the operator's actions. To generalize, recall that the model was trained with data from only two users, while it was evaluated against other fifteen users.

The remaining content of the paper is structured as follows. Section 2 provides an analysis of the current state-of-the-art related to the topic covered in this document. The data acquisition process and dataset specifications are introduced in Section 3. In Section 4, we explain the implementation, evaluation, and comparison of the two approaches to the force based operator's intent inference. The validation of the proposed system is presented in Section 5, and the conclusions and future work are discussed in Section 6.

2. Related Work

In this work, we are primarily interested in exploring force based industrial collaborative robotic tasks, that is those in which the physical interaction plays an essential role in the accomplishment of the task. In particular, it is of great interest for us to carry out a twofold research of: (a) applications where humans physically interact with robots; and (b) datasets containing force based information extracted from human–robot interaction scenarios.

In the literature, several works have presented applications where humans and robots physically interact. However, it is difficult to find recent works where, as in ours, the physical interaction plays a major role in the execution of a shared task. Indeed, in most of the cases, the force exchange between humans and robots is ignored or undesired. Hence, we analyzed two groups of works: (a) those in which the physical interaction is ignored or undesired; and (b) those in which the robot uses the force based information to adapt its state. Regarding the first group of works, Cherubini et al. [11] discussed a collaborative scenario where a human and a robot shared the task of Rzeppa homokinetic joint insertion. In this case, even though there was an exchange of force, unlike in our work, the robot just remained stiff and did not use the force based information to adapt its state. Maurtua et al. [12] described a set of experiments aimed at measuring the trust of workers on fenceless human–robot industrial collaborative applications. In all the experiments, the force was undesired; thus, the robot stopped when an external force was detected. De Gea Fernández et al. [13] described another industrial situation in which two robotic arms collaborated with an operator. The robots avoided the physical interaction with the human as long as possible, and when a physical interaction occurred, they remained in a compliant mode so that the force was ignored. Raiola et al. [14] addressed the problem of learning virtual guiding fixtures, analogous to the use of a rule when drawing, in human–robot collaboration. Even though there was physical interaction during the task execution, the robot did not use the force based information while guiding the human. In the work presented by Munzer et al. [15], a human and a robot performed sub-tasks of a shared task: wooden box assembling. The robot and the human shared forces, and the robot was able to adapt to the situation, but not using the force, just using vision, or being explicitly asked to do it by voice commands or instructions using a graphical interface. Some recent works presented cases in which robots adapted their behavior based on the physical interaction between humans and robots. Peternel et al. [16] proposed to estimate human fatigue to adapt how much a robot is helping in human–robot collaborative manipulation tasks: sawing and surface polishing. Rozo et al. [17] proposed a framework for a user to teach a robot collaborative skills from demonstrations. Specifically, they presented an approach that combined probabilistic learning, dynamical systems, and stiffness estimation, to encode the robot behavior along with the task. Hence, the method allowed a robot to learn not only trajectory following skills, but also impedance behaviors. Unlike in our work, in these two works, the adaptation was done at the low-level control of the robot by a hybrid force/impedance controller, while we did it at the symbolic level of the task. A scenario where a human and a robot physically interact through a handover of an object was discussed by Mazhar et al. [18]. Force signals were used to identify different phases of the sequence of actions. When a force threshold was exceeded, the system interpreted that the robotic hand should close to grasp the object during the handover. Zhao et al. [19] presented an operator's intention recognition approach inspired by a collaborative sealant task. The intentions, rather similar to ours, were also used to adapt the state of the robot, just as in our work. However, the interactions they proposed were simplistic as the classes could be discriminated between them with thresholds in the force. In our work, we recorded two different datasets, one that was similar to theirs, containing simpler mechanical movements, and another one that included more natural human–robot interactions. The latter was used during the experiments. Gaz et al. [20] presented a new robot control algorithm aimed at being used in a scenario where a robot grasps a piece while the operator polishes it. The proposed collaborative task was the same we used, but they considered only two robot modes: (a) stiffness, while the user polishes' and (b) compliance, while the user modifies the orientation of the end effector. Unlike in our work, there was no classification of the user's intentions; the force was directly applied to different parts of the robot: (a) a force sensor fastened to the robot's wrist; and (b) the rest of the robot's joints. Losey et al. [21] presented a comprehensive review of intent detection and other aspects within the context of shared control for physical human–robot interaction. Especially interesting was how this paper was structured, talking about three aspects covered in our work: (a) user intent recognition; (b) shared control between humans and robots; and (c) methods to inform the human operator about the robot's state.

In the literature, there are datasets extracted from robotics scenarios in which either the human–robot interaction is not physical or the force based tasks do not include interaction with humans. The former correspond to social robotics scenarios, where the most common means of interaction is not physical, but verbal. Those datasets usually contain video, speech (audio and transcripts), robot joint-sate, physiological data (e.g., bio-signals), or subjective data in the form of questionnaires [22–26]. On the other hand, it is possible to find some datasets containing force/torque data extracted from robotic scenarios in which robots and humans do not interact. Yu et al. [27] presented a dataset in the context of pushing tasks where a robot pushed an object along a specific surface. For each combination of an object's shape and a surface's material, these data contained forces in the pusher and poses of both the object and the pusher. Another interesting dataset involving forces was introduced by De Magistris et al. [28], where the authors presented a force-signal dataset used to learn peg-in-hole robot tasks. The dataset comprised force/torque and pose information for multiple variations of convex-shaped pegs. It was used to train a robot to insert polyhedral pegs into holes. Huang et al. [29] presented a dataset containing force/torque signals and poses of an end effector tool. Data were recorded from humans performing a set of different motions making use of the same tool that the robot would use, enabling the transference of knowledge. Datasets containing information about physical and force based human–robot interaction would be useful for collaborative robots to learn different task-dependent knowledge. Nevertheless, to the best of our knowledge, there is no available dataset containing force/torque data that comes from the physical human–robot interaction during a shared task.

3. Force Based Dataset of Physical Human–Robot Interaction

In this section, we provide all the relevant information related to the dataset (http://doi.org/10.5281/zenodo.3522205) used along the evaluations presented in this work. The dataset consisted of force/torque signals resulting from the physical human–robot interaction during the performance of a collaborative task, polishing a piece. The dataset was geared to teach robots to identify and predict humans' intentions during the proposed shared task. In the upcoming paragraphs, we first introduce the industrial collaborative scenario in which we used the dataset. Then, we explain the different sorts of operator intents we wanted to infer. Finally, we analyze the specifications of the dataset and how the data was collected. Note that we assume that the dataset was properly gathered and that it does not contain any outliers.

3.1. The Industrial Collaborative Robotic Scenario

In this work, we consider a realistic industrial scenario inspired by a manufacturing line of car emblems. We focus on one sub-process where the emblems are to be coated, and they must be totally clean and polished. Currently, the plant operator picks, inspects, and polishes the emblems, to finally place them into another location where they are coated. The objective is that a robot and the human share the task collaboratively. We have redesigned the process so the robot is in charge of the picking and placing tasks, while the operator still inspects and polishes the emblem. Once the robot posed the piece in front of the operator, the human could perform different actions over the emblem while the robot should infer those actions and adapt to them. In this scenario, the principle means of human–robot interaction was force based. The interaction should be natural for the human, and the reaction time of the robot should ensure a fluent and efficient collaboration. Note that it was not within the scope of this work to tackle how the robot grasps and places the emblems. Instead, we focused on how the robot, while offering the emblem, can infer the operator's intent and adapt its state appropriately.

3.2. Types of Operator Intents

Once the robot was offering the emblem to the user, we considered three different operator's intents: (a) polishing, (b) moving the robot, and (c) grabbing the object. Analogously, there were three different states of the robot w.r.t. them: (a) increasing stiffness (named "hold"), (b) decreasing stiffness ("move"), and (c) releasing the object ("open gripper"). In the first action, the operator should be able to do the main objective of the task, polishing the emblem. When applying this sort of force, the robot should be stiff. Otherwise, the polishing action would not succeed. The second operator's intent was regarding ergonomics in industrial scenarios. The operator could get tired of polishing the pieces in the same pose or there could be another operator with different corporal dimensions and/or abilities. Hence, this time, the force should be done to move the robot to a more comfortable pose. Finally, we also contemplated the case in which the human wanted to grab the object (emblem), pulling it from the robot's gripper. In this case, the robot should open the gripper to release the piece. These three actions should be performed naturally, and since they have a fundamental effect on the progress of the shared task, the robot should be able to react to them. It is worth mentioning that they were chosen considering the shared task from the scenario proposed in Section 3.1.

3.3. Dataset Specifications

The dataset was recorded using an ATI Multi-Axis Force/Torque Sensor Mini40 SI-20-1, which was fastened to the wrist of the robot, the basis of the end effector (see Figure 2a). We used the default configuration of the sensor, and the measurements were taken at a frequency of 500 Hz.

Every sample contained a single sort of interaction, from the beginning to the end of the physical contact. It is worth mentioning that the gathered data samples were not of the same length, ranging from half a second to three seconds long. In the dataset, the shorter samples were padded with zero values at the end of the temporal sequences so that all of them had the same length. The dataset contained six different files per each of the three classes, which corresponded to the six axes of the force sensor. Each file was named using the force/torque axis and the class label; hence, users could read the samples included in each file and label them appropriately.

Although we aimed to infer force based human intentions from natural and therefore ambiguous human–robot interactions, we first evaluated our method with less human based intentions, but more distinguishable mechanical interactions. The mechanical dataset was used as a baseline to check if the machine learning algorithms we studied could solve a simplified version of the problem we faced. Meanwhile, the natural dataset was employed to evaluate (see Section 4) and validate (see Section 5) the proposed approach to infer humans' intentions. In the mechanical dataset, each class followed distinct movement patterns, which produced completely different force signals. Therefore, the samples of each of the intentions/classes were distinguishable from each other. On the contrary, in the natural dataset, the movement patterns between classes were much more similar to each other; meaning there was more ambiguity among samples of different classes, which made classifying more complicated. In Section 4.5, we evaluate how the chosen machine learning approach (see Section 4.4) performed when it was individually trained and tested with each of the datasets.

Since it was expected to be easier to classify, the mechanical dataset only contained 600 samples. Recall that we had three classes, and we used two users; thus, each user performed 100 samples of each class. The physical contact was always done following restricted patterns for each intention/class. Figure 3 depicts both, the different axes in which the operator was supposed to apply the force and the corresponding force signals we detected using the sensor. For the polishing intention, we moved periodically only in the axis Y, and we pushed towards the robot, the negative Z-axis (Figure 3a). In order to move the robot, we moved just in one direction for each sample and only in the Y-axis (Figure 3b). Finally, to grab the object, we pulled the robot's end-effector towards ourselves, the positive Z-axis (Figure 3c).

Unlike with the mechanical dataset, the natural dataset contained more samples, 900. Recall that we had three classes, and we used two users; thus, each user performed 150 samples of each class. In this case, the physical contact for each intention/class could be done following several natural patterns, which increased the ambiguity between classes. In Figure 4, it is possible to see the different axes in which the operator was supposed to apply the force and the corresponding force signals we detected using the sensor. For instance, the intention of polishing could now be done by describing circles and also using the X-axis (Figure 4a). The patterns to move the robot now included any of the directions of the three spatial axes (Figure 4b). Finally, the operator could now try to grab the object pulling, but not only towards the exact direction of the Z-axis (Figure 4c).

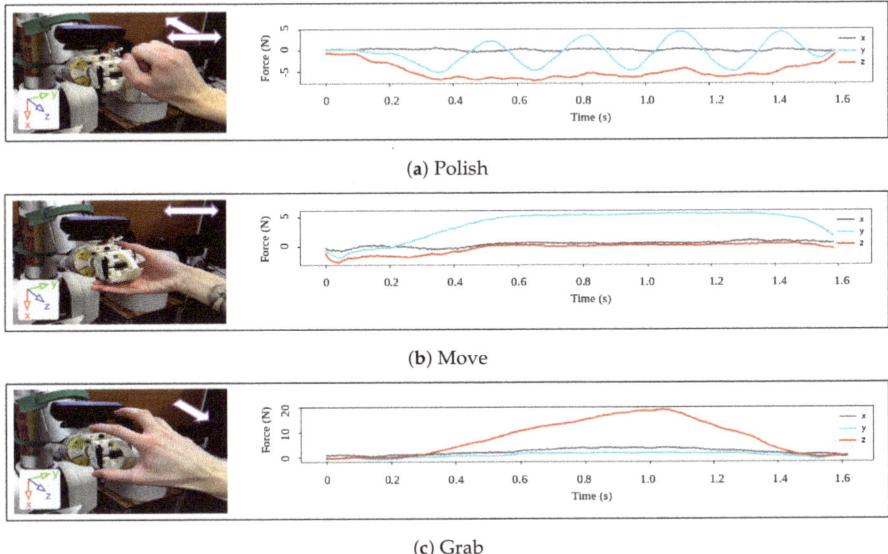

(a) Polish

(b) Move

(c) Grab

Figure 3. Mechanical dataset. Human movement patterns (left side) and appearance of the force signals produced by those patterns (right side). Observe how each class (**a–c**) is quite distinguishable from the rest even after only 0.4 s. Making use of this dataset to train a model would allow predicting fast with enough confidence. Nevertheless, the movement patterns of the user would be too restricted, and the human–robot interaction would not be natural.

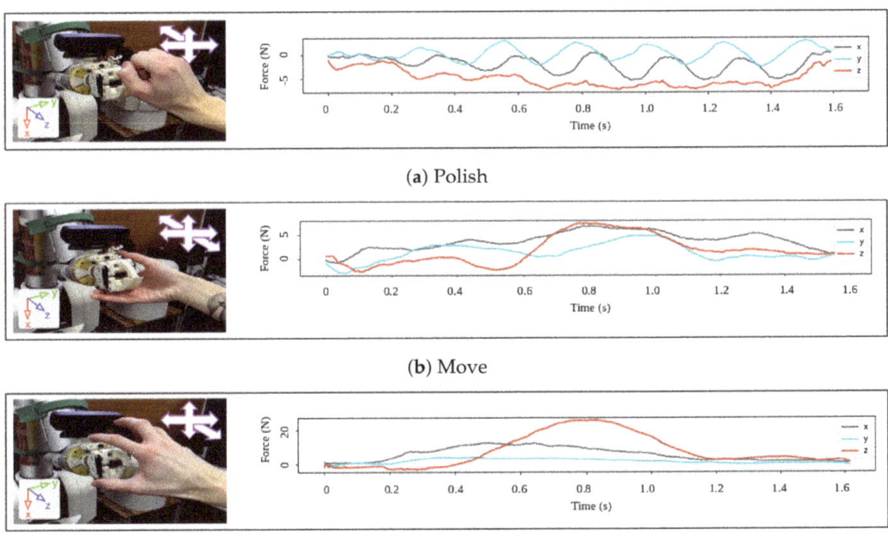

(a) Polish

(b) Move

(c) Grab

Figure 4. Natural dataset. Human movement patterns (left side) and appearance of the force signals produced by those patterns (right side). Observe how each class (**a–c**) is still similar to the rest even after 0.4 s. Due to the richness in movements, a model trained with this dataset would allow a natural human–robot interaction.

It is worth discussing the visual differences between the signals of both datasets. In the mechanical dataset, signal forces looked different when we considered the entire time series, but also after 0.4 s of signals. Forces occurred in isolated axes for each of the operator's actions/intents, and ambiguity between classes was kept to a minimum. Hence, it was possible to discriminate between classes with a reduced amount of force information. This was not the case for the natural dataset. Of course, signals from different classes were still distinct if we considered the whole temporal sequence. Nevertheless, unlike with the mechanical dataset, we could not be so sure about the label of each of the signals after only 0.4 s. Please, recall that, although for illustrative purposes, the figures only show the linear forces, our classification process used both torque and linear signals. Together with the dataset, we also provide some Python code to run our proposed approaches and use the data (http://doi.org/10.5281/zenodo.3522205). Therefore, other people can learn how to use the dataset on their own.

4. Force Based Operator's Intention Inference

In order to infer humans' intents, we have evaluated the performance of two approaches using the natural dataset. We compared them and chose one, which was used during the validation carried out in Section 5. Finally, the chosen approach was also used to analyze the differences between the natural and mechanic datasets. These results are part of the experimental findings presented in our work. One of the approaches, kNN + DTW, was based on a classifier that directly used the raw sensor data to perform the inference, whereas the other one, GPLVM + SVM, used a lower dimensional representation of the data. Recall that we sought a natural human–robot interaction, a fast reaction of the robot, and if possible, an approach that dealt with heterogeneous industrial contextual data.

4.1. Evaluation Setup for the Proposed Approaches

The performance of the proposed approaches was evaluated following the considerations explained in this section. Cross-validation without replacement was applied ten times, and the data were randomly split into training (75%) and test (25%) sets. The chosen metric to evaluate the performance was the F1-score, which captures both the precision and the recall of the test.

In order to fulfill the requirement of a profitable robot reaction, the prediction time should be short enough so that the proposed methods apply to our realistic scenario. For that reason, we did not consider all the samples, but smaller portions of them (windows), which contained only their initial information. In total, five different window's sizes were evaluated: 0.1, 0.2, 0.5, 0.7, and 1 s (see Figure 5). The intuition is that the larger the sampling window, the higher would be the chances to classify the human's intention properly, but the longer the operator would need to wait until the robot reacts to the interaction. Therefore, we aimed to find a trade-off between the prediction time and the classification performance. Our experience said that 1 s was a convenient amount of prediction time for an efficient and feasible human–robot collaboration. Thus, longer inference time would be undesirable. Note that the total prediction time would include both the sampling window's size and the time the approach needs to infer the label of the sample.

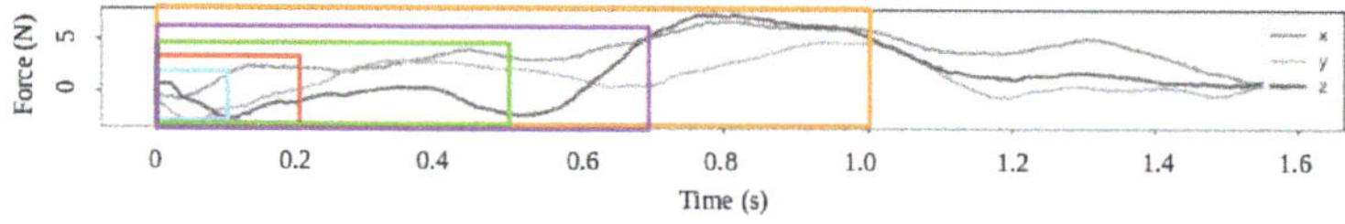

Figure 5. Sampling windows evaluated to find an optimal classification-reaction time ratio. The windows correspond to: 0.1 s (cyan), 0.2 s (red), 0.5 s (green), 0.7 s (purple), and 1 s (orange). Recall that one second is our task limit time for achieving a suitable human–robot interaction.

4.2. Raw Data Based Classification

In this approach, using the data obtained from the sensor directly, the classification was done utilizing a k-Nearest Neighbors (kNN) classifier with Dynamic Time Warping (DTW) [30] as the metric.

In particular, we used $k = 1$. While being a simple method, 1NN + DTW's performance seems to be hard to beat by other approaches in time series classification problems [31].

4.2.1. Implementation Details of the Raw Data Based Classification

Dynamic time warping is a time dependent algorithm used to measure similarity between two temporal sequences that may vary in speed. For instance, similarities in polishing could be detected using DTW, even if the operator polishes faster or slower than on other occasions. DTW is a computationally-intense technique, with quadratic time and memory complexity. However, there are some ways to accelerate computation. In our case, we used the library Fast DTW [32]. DTW is meant to be utilized for univariate time series, which was not our case since we had six sensor axes. From the literature, we know at least two obvious approaches to tackle this and generalize DTW for multi-dimensional time series: dependent and independent DTW (see Figure 6) [33]. The kNN classifier was taken from the scikit learn library [34]. Since default implementations of both kNN and Fast DTW do not allow working with multi-dimensional time series, it was necessary to adapt the libraries we used. Apart from those modifications, we used the values set by default.

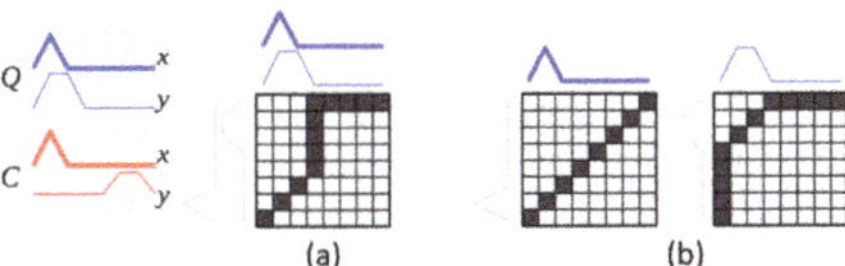

Figure 6. Dynamic Time Warping (DTW) for multi-dimensional time series: dependent (**a**) and independent (**b**) DTW. The former consists of computing the DTW similarity path of both dimensions (axis) at the same time. The latter is much simpler; normal DTW is computed separately on each dimension and their results added subsequently.

4.2.2. Evaluation of the Raw Data Based Classification

The proposed method, 1NN + DTW, was evaluated for each of the window sizes previously defined, concerning the classification performance and the inference time per sample. Recall that two different implementations of multi-variate DTW were used, dependent and independent, DTWd and DTWi, respectively. Due to the lazy learning nature of the kNN classifier, we also evaluated how the length of the samples fed to the classifier affected the inference time. In particular, we sub-sampled the measurements of the windows to smaller portions. We considered five different lengths, which were expressed as the percentage of the window's length that remained after the sub-sampling: 100% (no sub-sampling), 8%, 6%, 4%, and 2%. Figures 7 and 8 show the results of the evaluation.

There are many conclusions that could be drawn by analyzing the information shown in Figures 7 and 8. In the first place, the bigger the window, the better the performance; see the evolution of F1-score in Figure 7. It is also true that the growth of the window's size resulted in an increment of the inference time per sample (see Figure 8). This is reasonable since the kNN algorithm is a lazy learner. Any time a new sample is to be classified, the similarity between that sample and the rest of the training samples is computed. Hence, the longer the samples, the more time it takes to compute the similarity, prolonging the whole inference process.

The best F1-score result (99.24%) was obtained for the case of using DTWd with the window size of one second and sub-sampling of 6% of the total window's size (see the orange bar in Figure 7). The inference time per sample for this same case was above half a second (0.7 s), which can be seen looking at the same bar in Figure 8. Therefore, the total operator's intent inference time would be around 1.7 s, which is above the one second we sought, so this was not a valid alternative.

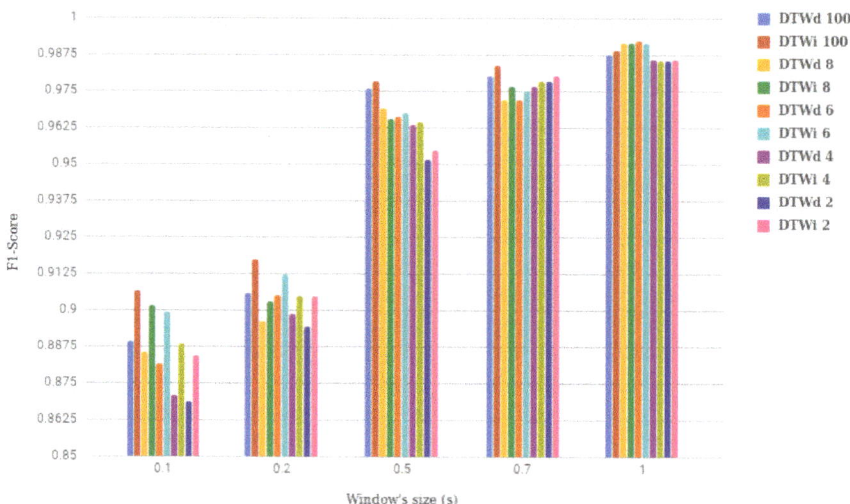

Figure 7. F1-score values for the different types of raw data based classification (dependent and independent DTW), sampling window's size (0.1, 0.2, 0.5, 0.7, and 1.0 s), and percentage of sub-sampling (where 100 means non-sub-sampling). The longer the sampling window, the better the classification performance. Observe how, for our task, a 0.5 s sampling window already provided a very good F1-score.

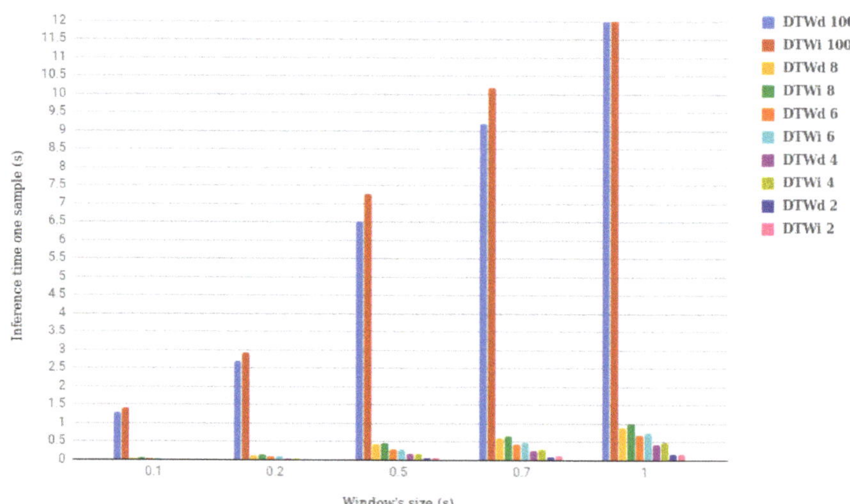

Figure 8. Graphical representation of the values of the inference time per sample for the different types of raw data based classification (dependent and independent DTW), sampling window's size (0.1, 0.2, 0.5, 0.7, and 1.0 s) and percentage of sub-sampling (where 100 means non-sub-sampling). The longer the window of data we consider, the longer the inference time. The total time to recognize the operator's intent is the addition of the window's size (horizontal axis) plus the inference time per sample (vertical axis).

Fortunately, reducing the window's size, while helping to reduce the inference time, did not decrement the performance too much. As can be seen in Figure 7, from windows bigger than 0.5 s, the value of the F1-score was always above 95%. The best F1-score value for that window was around 97.5%, which was a really good result. It corresponded to the case of using all the data within the window's size together with DTWi (red bar). Nevertheless, if we used that configuration for the approach, the time needed to infer the operator's intent would be above seven seconds, once again undesirable.

We needed to find the most convenient combination of: the DTW version, sampling window size, and whether sub-sampling was needed or not. Indubitably, we discarded the case in which sub-sampling was not applied, since the inference time (blue and red bars in Figure 8) was always above the desired one. Any case that used the one second window could also be dismissed, since the performance was not much better than for the case of using 0.5 or 0.7 s windows. Hence, we focused on the 0.5 and 0.7 s windows, in which there was not any combination that, at the same time, performed better and faster than the rest. Nonetheless, should we choose one case, we would select a case in which the trade-off between inference time (0.8 s) and performance (97.99%) was rather good. This case corresponded to DTWi, a window of 0.7 s, and sub-sampling of the data to 2% of the window's size (pink bar in Figure 7).

4.3. Feature Based Classification

In this section, we propose a twofold machine learning approach to infer the human operator's intentions. First, we reduced the dimensionality of the data using an unsupervised method: Gaussian Process Latent Variable Model (GPLVM) [35]. Then, we used a Support Vector Machine (SVM) classifier, which was trained using the lower dimensional representation of the data. GPLVM is a non-linear dimensionality reduction method that can be considered as a multiple-output GP regression model where only the output data are given. The inputs are unobserved and treated as latent variables; however, instead of integrating out the latent variables, they are optimized. By doing this, the model gets more tractable, and some theoretical grounding for the approach is given by the fact that the model can be seen as a non-linear extension of the linear Probabilistic PCA (PPCA) [36]. Note that in this case, the temporal sequences are just considered as long feature vectors, so that the temporal relation between subsequent signal measurements is not explicitly considered. However, dimensionality reduction has proven to be an effective technique in time series analysis, in which data are remarkably high dimensional [37–39].

4.3.1. Implementation Details of the Feature Based Classification

The implementation of the proposed method, GPLVM + SVM, relied on two existing libraries: the GPy library [40] for the dimensionality reduction and the scikit learn library for the SVM classifier [41]. In the case of the latter, we used the default values for all the parameters. However, concerning GPLVM, it was necessary to set some parameters: kernel, optimizer, and the maximum number of optimization steps. Firstly, we chose a kernel that was a combination of the Radial Basis Function (RBF) kernel together with a bias kernel. The RBF kernel was selected because it is one of the most well known kernels for non-linear problems. We added the bias kernel to enable the kernel function to be computed not only in the origin of coordinates. Secondly, for the optimization process, we used one of the optimizers already implemented in GPy, limited-memory Broyden–Fletcher–Goldfarb–Shannon (BFGS) [42]. We chose this optimizer because, unlike others included in the library, it was quite stable concerning the number of optimization steps needed to converge. Finally, the maximum number of optimization steps was set to 5000, which in most cases was enough for the optimization to converge.

The implementation of the GPLVM algorithm allowed us to use two different types of latent variable inference: with the optimization step (GPLVM-op) and without the optimization step (GPLVM). For us, the most relevant difference between them was that the inference with optimization took more time, but it would be more correct in theory and would lead to more accurate results. Nevertheless, as we will see in Section 4.3.2, the inference with optimization did not always ensure better performance. Once an already optimized GPLVM received a new sample to infer its latent variables, the global inference process was divided into three steps. The first step, nearest neighbor search, was focused on finding which of the training samples was the most similar to the new sample. This was done by computing the similarity between the new sample and all the training samples employing the Euclidean distance. The second step, latent variables' initialization, consisted of setting the value of the inferred latent variables to the values of the latent variables of the nearest neighbor found in the previous step. Finally, during the third step, latent variables' optimization, the value of the initialized latent variables was refined through optimization. Figure 9 depicts the global pipeline of the inference process detailed above.

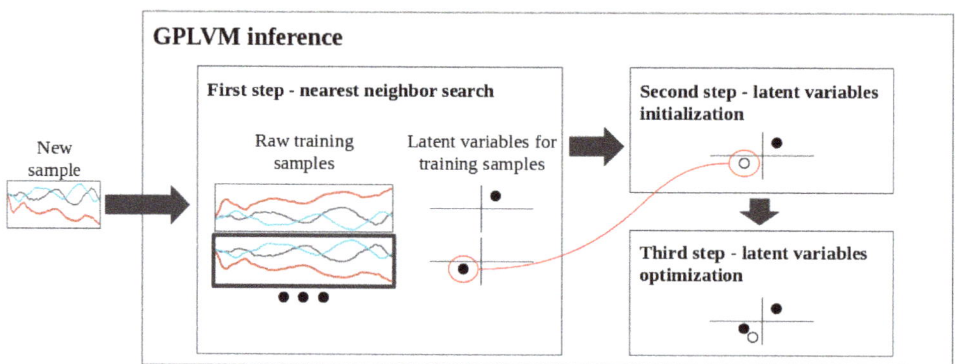

Figure 9. Global GPLVM inference process of the latent variables given a new sample in the higher dimensional space. First, the most similar training sample to the new sample is found using Euclidean distance. Second, the value of the latent variables of the most similar training sample (black dot in the first step) is used to initialize the inferred value (see the white dot in the second step). Third, the GPLVM model is optimized considering the new sample, which results in a refinement of the inferred latent variables. GPLVM with optimization includes the three steps; GPLVM without optimization stops after the second.

4.3.2. Evaluation of the Feature Based Classification

The proposed method, GPLVM + SVM, was evaluated for all the different already mentioned window sizes about both the classification performance and the inference time per sample. A priori, we did not know which size of the latent space would produce a good performance. Therefore, different sizes of latent space were also evaluated: 2, 3, 5, 10, and 20 latent variables. Besides, the two types of GPLVM were evaluated as well: optimized (GPLVM-op) and non-optimized (GPLVM). Figure 9 depicts the global modular structure of the GPLVM inference process. Figures 10 and 11 show respectively the results of both the F1-score and the inference time with respect to the different window sizes and the GPLVM methods used.

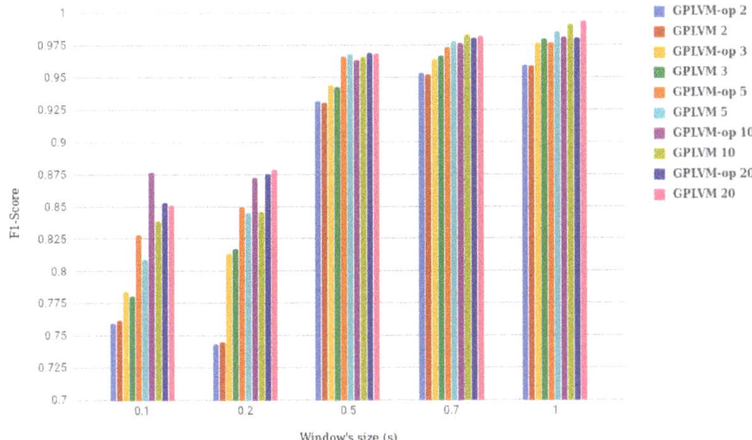

Figure 10. F1-score values for the different types of feature based classification (optimized (op) and non-optimized GPLVM inference), sampling window's size (0.1, 0.2, 0.5, 0.7, and 1.0 s), and number of latent variables (2, 3, 5, 10, and 20). Note that the bigger the number of latent variables, the better is the result, which also happens with the window size. Furthermore, observe that in some cases where the window's size is very small (0.1 and 0.2 s), the shorter window outperforms the longer one by a small amount. This behavior is counter-intuitive, but possible due to the still negligible information contained within those small samples and the random selection of the training set.

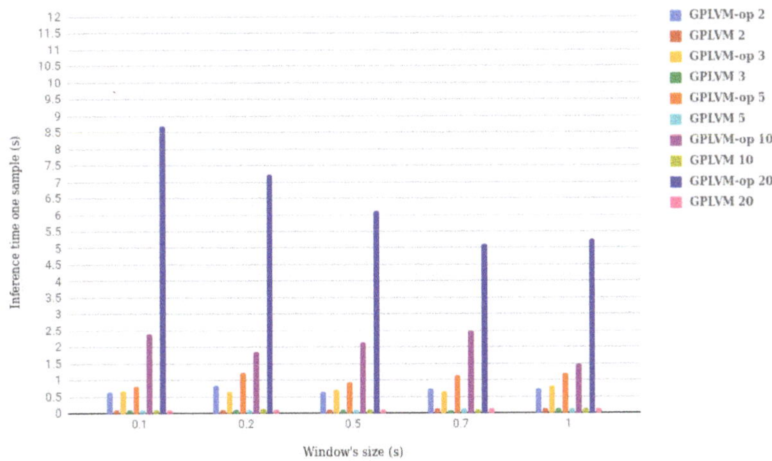

Figure 11. Graphical representation of the inference time per sample for the different types of feature based classification (optimized and non-optimized GPLVM inference), sampling window's size (0.1, 0.2, 0.5, 0.7, and 1.0 s), and number of latent variables (2, 3, 5, 10, and 20). GPLVM-op leads to longer inference time than GPLVM, which also applies when the number of latent variables grows.

Evaluating in detail the results depicted in such figures, probably, the most evident conclusion is the effect of the optimization during the inference step in the GPLVM. The inference time per sample was always longer when GPLVM inference was optimized. Indeed, that time grew accordingly to the number of latent variables (see Figure 11). Another interesting finding was that the inference time per sample, when there was no optimization, remained quite short and stable no matter the window's size nor the number of latent variables (see Figure 11). Hence, in terms of inference time, GPLVM without optimization was preferred. Moreover, as can be seen in Figure 10, the performance score between both optimized and not optimized versions was negligible. This fact reinforced the previous result, allowing us to conclude that the non-optimized version of GPLVM was the most convenient alternative.

Focusing on Figure 10, it is observable that the more latent variables we used, the better was the result. Specifically, for the cases in which we used two and three latent variables (specially two), the performance (F1-score) was usually much poorer. The best result in terms of performance, an F1-score of 99.33%, corresponded to the GPLVM version without optimization, the window of 1 s, and 20 latent variables. The inference time per sample was around 0.15 s, so the total inference time was 1.15 s, slightly superior to the one second we set as desirable. Thus, we decided to reduce the window's size to 0.7 s. In this case, the best alternative was to use 10 latent variables and, again, the non-optimized GPLVM. This resulted in losing a bit of quality in the performance, from 99.33% to 98.14%, not noteworthy, but decreasing the time from 1.15 to 0.85 s, fulfilling our requirements.

4.4. Raw Data Based vs. Feature Based Classification

In this section, we compare only the best combination of parameters for each of the two studied methods. Finally, we selected one of them to be used during the experimental validation proposed in Section 5. Recall that at the beginning of this work, we stated some requirements that the selected approach should fulfill. The human and the robot should interact naturally, and the robot adaptation should last one second at most. Furthermore, in the future, we aim to consider the contextual information of the industrial processes surrounding the proposed collaborative task. Hence, it would be desirable that the method to infer the human's intention could deal with heterogeneous data, not only temporal sequences.

The selected combination in the case of 1NN + DTW ensured an inference time of 0.8 s and a performance score of 97.99%, which was rather good. It corresponded to using independent DTW, a window of 0.7 s, and sub-sampling of the data to 2% of the window's size (see Section 4.2.2 for more detail). When using GPLVM + SVM, the selection was GPLVM without optimization, a window of 0.7 s, and 10 latent variables. This approach resulted in an F1-score of 98.14% and an inference time of 0.85 s (see Section 4.3.2 for more detail). As we can see, the quantitative differences between the two alternatives were negligible. Therefore, to provide more useful insights into the comparison between 1NN + DTW and GPLVM + SVM, we analyzed them using more qualitative measures. They were extracted from the hands-on experience acquired along the developed work and were meant to ease the selection procedure.

- Ease of implementation: Both methods were relatively simple to implement and use. Conceptually and algorithmically, 1NN + DTW was a simple machine learning technique; only the versions of DTW for multivariate data presented a bit of difficulty. GPLVM was theoretically more complex, and reaching a profound understanding of the mathematical background of this technique would require effort. However, the GPy library eased the use of GPLVM without the need to dig too much into the theoretical details.
- Data visualization: GPLVM allowed us to project the sequential data samples into just a few latent variables and then visualize the data distribution in either 2D or 3D. This can be useful to analyze the dataset easily, and it was something that could not be done using 1NN + DTW.
- Generalization to other scenarios: This aspect is rather important for us because in the future, we would like to include heterogeneous environmental variables in the learning pipeline. Examples of contextual variables are: if the grasped object is heavy or not and if the user is inside the workspace or not. In this case, these two variables are binary and could be added to the feature vector of each sample to learn some environmental aspects related to safety. GPLVM could be used to reduce the dimensionality of temporal sequences to just a few features. Then, other contextual variables could be concatenated to the resulted feature vector, and SVM would

be used to learn not only the physical interactions but also the contextual information. 1NN + DTW, however, cannot deal with other data apart from sequential. It would be necessary to use a second kNN model with another metric (e.g., Euclidean) and then apply ensemble learning techniques.

Based on the previous analysis, we selected GPLVM + SVM. In particular, we proposed to use GPLVM without optimization during the inference, a sampling window of 0.7 s, and 10 latent variables. The first reason was that we thought GPLVM's generalization capabilities could help us in future works. In robotics, especially in industrial environments, data are presented in heterogeneous ways: sequential data, digital, etc. Let us consider one of the examples proposed in the generalization paragraph. If the object the robot grasps is too heavy, we could just add a "1" to the feature vector of latent variables and train the SVM classifier with the new extended vector. Therefore, it could be learned that even when the inferred human's intention is grabbing the object, the robot must never open the gripper if the object is too heavy. Of course, if we consider only one environmental variable, the easiest way to tackle this event would be to add a conditional statement to the control code of the robot. However, if the number of those variables increases, machine learning methods could help. Furthermore, GPLVM allowed us to visualize the distribution of the data we worked with, which could be especially useful if the dataset were enlarged by other people, and we wanted to see how the different datasets related to each other.

4.5. Comparison of Natural and Mechanical Datasets

In this section, we evaluate and compare the performance of the chosen approach, GPLVM + SVM, using both datasets, the natural and the mechanical. We assumed that the mechanical dataset would show a good performance even with a small sampling window sizes. Given that, we wanted to analyze if the proposed method, for the sampling window of 0.7 s, could work similarly well, not only with the mechanical, but also with the natural dataset. Recall that we chose to use the non-optimized GPLVM inference and 10 latent variables. Although the selected sampling window's size was 0.7, during this section, we tested the approach against the usual five sizes we used along the rest of the document. As was done previously, we used cross-validation without replacement ten times, and the data were randomly split into training (75%) and test (25%) sets.

Figure 12 depicts the F1-score values obtained from the evaluation of GPLVM + SVM against both datasets. This bar diagram shows that indeed, our previous assumption was true. In general, using the mechanical dataset, we obtained better results than with the natural data. Specifically, when the window's size was 0.2 s, the F1-score was even close to 95%. However, we also observed that for the window chosen for our validation with users, 0.7 s, the differences between the performance using any of the datasets were minimal. Therefore, the proposed approach worked quite well even when the dataset contained more natural samples of physical human–robot interaction.

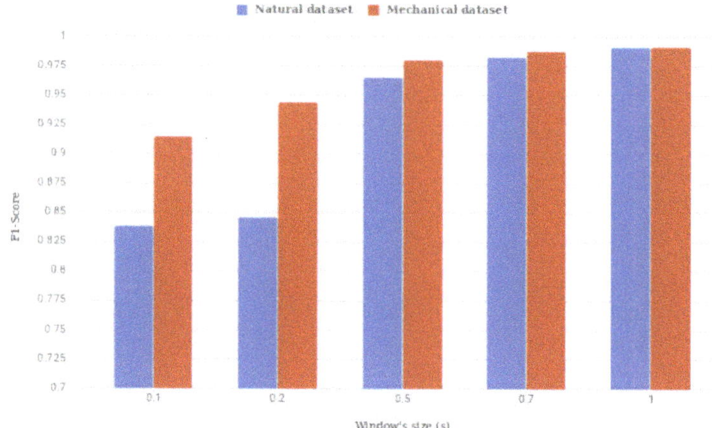

Figure 12. Evaluation of the natural (blue) and the mechanical (red) datasets of the approach GPLVM + SVM without optimization and 10 latent variables. The mechanical data need less force information to classify with good quality. However, if the window of the force signal is large enough (more than 0.5 s), the model behaves similarly no matter whether the data are mechanical or natural.

5. Validation: Inferring Operator's Intent in a Realistic Scenario

To validate the selected approach, GPLVM + SVM, we set up an experiment in which several users individually collaborated with a robotic arm according to the industrial scenario of polishing car emblems. The validation was conducted using fifteen healthy individuals within an age range of 18 to 35. Users were selected among people who had knowledge about the robotics domain and had been in contact with robots before. We did not include people with reduced mobility or any cognitive disability, which could affect the perception of the robot's behavior, endangering the users' integrity. Each of the users received an individual explanation, no more than five minutes, about how they were expected to interact with the robot. This included both general information about the system and particular notions about the expected movements for each of the three classes/intentions. Nevertheless, the users were not allowed to train before the evaluation began, because we wanted to evaluate if there was an adaptation of the user to how the system inferred the different intentions. Users were also informed about their rights, possible risks, and were asked to sign an ethical approval specifically designed for this experiment. Note that we followed CSIC (Spanish National Research Council) ethical procedures and asked for ethical consent from the Human Subject Research Committee of CSIC before the validation was conducted.

Recall that the parameter combination for the chosen approach was: GPLVM without optimization, a sampling window of 0.7 s, and 10 latent variables. In this section, we give the flavor of the validation setup, and we evaluate and discuss the obtained results.

5.1. Setup

The validation setup was aimed at fulfilling the needs required by a human and a robot to collaborate on an industrial task in which the force exchange is not only present, but fundamental for the accomplishment of the task. Using the force based information, the robot should be able to identify the intent of the operator (Section 3.2) and to adapt its state/behavior to it. In order to provide a bi-directional communication, we equipped the robot with a force sensor, used to measure the interaction from the human to the robot, and an armband made of LEDs through which the robot informed the user of its internal state. The latter allowed us to display different patterns (see Figure 13). The finite state machine of the control of robot during the validation experiment is shown in Algorithm 1.

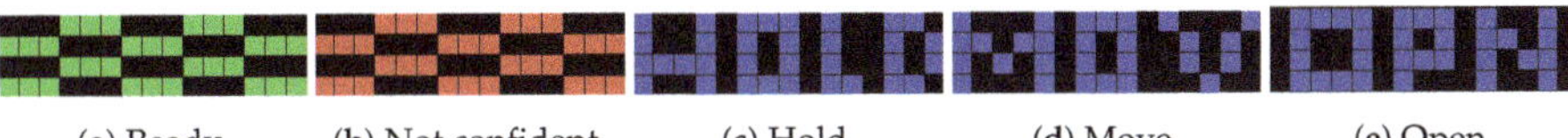

(**a**) Ready.   (**b**) Not confident.   (**c**) Hold.   (**d**) Move.   (**e**) Open.

Figure 13. LED patterns used by the robot to communicate with the user using the robot's armband. (**a**) Green pattern used to indicate when the robot is ready for physical interaction. (**b**) Red pattern indicating low classification confidence (<70%). Textual patterns showing the state of the robot when user intents are identified with high confidence: (**c**) "hold" (polish intent), (**d**) "move" (move intent), and (**e**) "open" (grab intent). The character "e" could not be expressed due to the four row armband matrix restriction.

Algorithm 1: Finite state machine of the control of the robot during the validation.

Data: Force sensor's signals
Result: Robot's state adaptation

```
 1  initialization;
 2  while true do
 3  │   robot in initial pose;
 4  │   inform operator: robot is ready for interaction;
 5  │   wait for physical contact;
 6  │   if detected physical contact then
 7  │   │   prepare sample from raw sensor data;
 8  │   │   infer operator's intention;
 9  │   │   if inference's confidence ≥ 0.7 then
10  │   │   │   inform operator: next robot's state;
11  │   │   │   adapt robot's state to the inferred intention;
12  │   │   else
13  │   │   │   inform operator: the inference's confidence was low;
14  │   │   end
15  │   else
16  │   │   do nothing;
17  │   end
18  end
```

Recall that this scenario was inspired by a real industrial case in which an operator was meant to inspect and polish car emblems. Please refer to Figure 2a to see the different parts of the robot setup used. We can only show the adapter where the emblem is attached since emblems contain private commercial brand logos and cannot be shown due to confidentiality agreements. Another important aspect related to the setup is how the user is located with respect to the robot. We chose to pose the operator in front of the robot so that the physical interaction was comfortable. During the experiment, the operator will have a rag that would be used to polish. Figure 2b shows an example of the pose of a user while polishing. A video of the validation with users can be found at www.iri.upc.edu/groups/perception/SIMBIOTS.

5.2. Evaluation

Each user was asked to perform thirty trials randomly selected from the three operator's intent/actions explained in Section 3.2. We made sure that among the thirty trials, ten corresponded to each of the three classes/intentions. Note that since trials were randomly arranged for each person, there could not be any bias in our evaluation due to the order of the trials. Both the ground truth and the inferred value were annotated for each user's trial. In this section, we analyze the overall performance of the system (confusion matrix) and the overall adaptation of the users throughout the experimental validation.

A confusion matrix of the performance of the system for each user was computed, then we calculated the final mean confusion matrix shown in Figure 14, which contained the average result for all users. The most obvious observation one can make is that the "move" intent was the easiest to identify. Indeed, the confusion matrix was not symmetric, and this class showed a large percentage of false positives, which was a symptom of a clear bias of the model in favor of this class. This can be better understood by looking at Figure 15. This figure shows the sample distribution in the three-dimensional space defined by the most significant/discriminating latent variables among the ten used. We can observe how the samples from the "move" class fell in the middle of the other two classes, which explains why there were many false positives, shared with the other two classes. However, given the bias in favor of this class and the higher proximity to the "grab" class, this latter was the class with the biggest number of false positives.

As stated before, we also studied if there was an adaptation of the users to the system, which would be observable in the performance of the system along the validation experiments. Recall that users only received a short explanation of the three classes and in which axes they could perform the movements for each action. There was ambiguity among classes, and users had a particular way to move for each action. Because of this, during the first trials, the system's performance was poorer. When we talk of adaptation, we mean that the users understand which movements for each class ensure a better performance of the system. Note that this is possible because users could see the result of the inference.

	Grab	Move	Polish
Grab	0.6133	0.3800	0.0067
Move	0.1200	0.8667	0.0133
Polish	0.0667	0.1667	0.7667

Figure 14. Normalized confusion matrix of the performance of the system during the validation with all users and trials. The matrix is non-symmetric, and the biggest portion of misclassified samples of the classes "grab" and "polish" are inferred as "move", which indicates the existence of some bias in favor of the class "move".

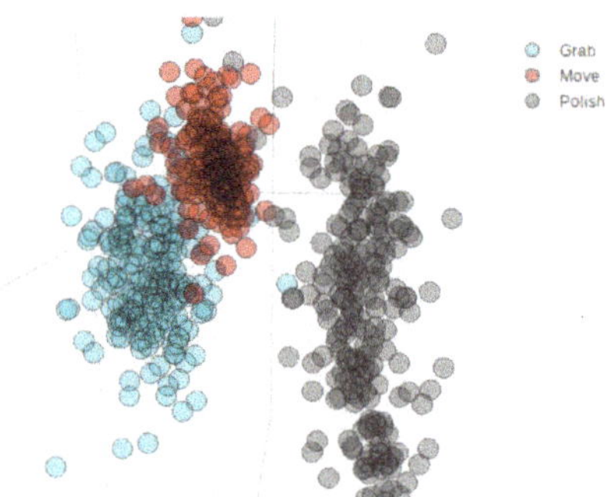

Figure 15. Single perspective of the data visualization using the three most discriminating latent variables from the original ten. The distribution of the data in this lower space shows that the samples of the class "move" are rather close to the other two classes, which could be the reason why the model seems to be a bit biased in favor of this class.

We computed the average performance of the system for all the trials and users, and the result showed a positive slope of the trend line for the F1-score (Figure 16). We considered that once the trend line was above 0.8, users had already adapted. In our case, this corresponded to the last five trials of the experiment.

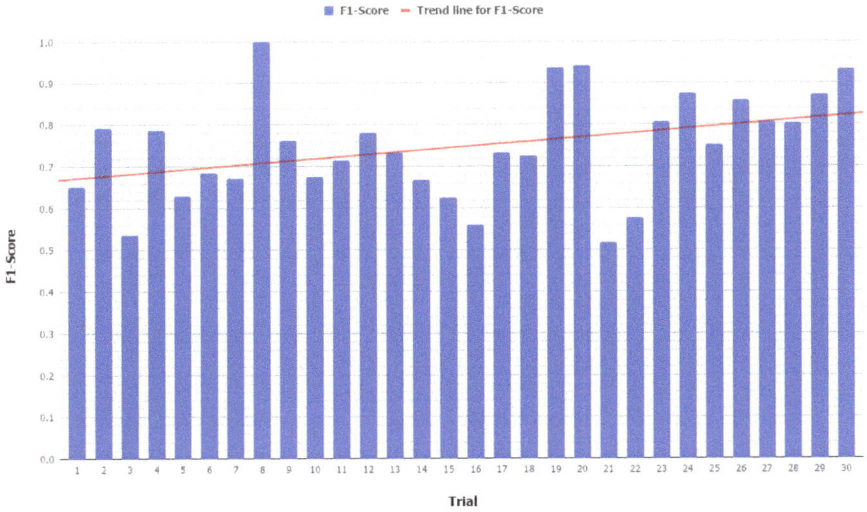

Figure 16. Average F1-score of the system for all the users along with the experiment's trials. The positive slope of the trend line for the F1-score is an indicator of the adaptation of the users to the system. Please recall that none of the users followed the same sequential trial set since they were randomly generated.

6. Conclusions

In this article, we presented our work on inferring operators' intent throughout the execution of an industrial collaborative task in which a robot and an operator exchanged forces while sharing the accomplishment of the task. This work consisted of three major contributions: (a) force based operator's intention inference; (b) force based dataset of physical human–robot interaction; and (c) validation of the whole system in a scenario inspired by a realistic industrial application. In our work, the physical interaction between the robot and the human not only existed, but also played a major role since it was the main source of information for the robot to infer the human's intent. Were humans and robots to collaborate in industrial environments in the factories of the future, the main interaction would be physical. Hence, our work means a step forward to enhance humans' and robots' collaboration in real case studies with more natural and user-friendly interaction. In the future, we will consider exploring other model based representations of the inherent contextual knowledge of collaborative shared tasks, to extend our current system to a wider range of more complicated scenarios.

Author Contributions: Data curation, A.O.-A.; software, A.O.-A.; supervision, S.F. and G.A.; Writing—review & editing, A.O.-A., S.F. and G.A.

Funding: This work is supported by the Regional Catalan Agency ACCIÓthrough the RIS3CAT2016 project SIMBIOTS(COMRDI16-1-0017) and the Spanish State Research Agency through the María de Maeztu Seal of Excellence to IRI (Institut de Robòtica i Informàtica Industrial)(MDM-2016-0656) and the HuMoUR project TIN2017-90086-R (AEI/FEDER, UE).

Acknowledgments: Firstly, this work was possible thanks to the help of the users that were part of the validation phase, so we thank all of them for their contribution. We cannot forget the help provided by Marc Maceira, who was fundamental throughout the dataset gathering and also during the experiment with users.

Conflicts of Interest: The authors declare no conflict of interest.

References

1. Michalos, G.; Makris, S.; Tsarouchi, P.; Guasch, T.; Kontovrakis, D.; Chryssolouris, G. Design considerations for safe human–robot collaborative workplaces. *Procedia CIRP* **2015**, *37*, 248–253. [CrossRef]
2. Villani, V.; Pini, F.; Leali, F.; Secchi, C. Survey on human–robot collaboration in industrial settings: Safety, intuitive interfaces and applications. *Mechatronics* **2018**, *55*, 248–266. [CrossRef]
3. Michalos, G.; Makris, S.; Spiliotopoulos, J.; Misios, I.; Tsarouchi, P.; Chryssolouris, G. ROBO-PARTNER: Seamless human–robot cooperation for intelligent, flexible and safe operations in the assembly factories of the future. *Procedia CIRP* **2014**, *23*, 71–76. [CrossRef]
4. Tsarouchi, P.; Michalos, G.; Makris, S.; Athanasatos, T.; Dimoulas, K.; Chryssolouris, G. On a human–robot workplace design and task allocation system. *Int. J. Comput. Integr. Manuf.* **2017**, *30*, 1272–1279. [CrossRef]
5. Wang, L.; Gao, R.; Váncza, J.; Krüger, J.; Wang, X.V.; Makris, S.; Chryssolouris, G. Symbiotic human–robot collaborative assembly. *CIRP Ann.* **2019**, *68*, 701–726. [CrossRef]
6. Roy, S.; Edan, Y. Investigating joint-action in short-cycle repetitive handover tasks: The role of giver versus receiver and its implications for human–robot collaborative system design. *Int. J. Soc. Robot.* **2018**, 1–16. [CrossRef]
7. Someshwar, R.; Edan, Y. Givers & receivers perceive handover tasks differently: Implications for human–robot collaborative system design. *arXiv* **2017**, arXiv:1708.06207.
8. Bauer, W.; Bender, M.; Braun, M.; Rally, P.; Scholtz, O. *Lightweight Robots in Manual Assembly—Best to Start Simply. Examining Companies' Initial Experiences with Lightweight Robots*; Fraunhofer IAO: Stuttgart, Germany, 2016.
9. Someshwar, R.; Meyer, J.; Edan, Y. A timing control model for hr synchronization. *IFAC Proc. Vol.* **2012**, *45*, 698–703. [CrossRef]
10. Someshwar, R.; Kerner, Y. Optimization of waiting time in HR coordination. In Proceedings of the 2013 IEEE International Conference on Systems, Man, and Cybernetics, Manchester, UK, 13–16 October 2013; IEEE: Piscataway, NJ, USA, 2013; pp. 1918–1923.
11. Cherubini, A.; Passama, R.; Crosnier, A.; Lasnier, A.; Fraisse, P. Collaborative manufacturing with physical human–robot interaction. *Robot. Comput. Integr. Manuf.* **2016**, *40*, 1–13. [CrossRef]
12. Maurtua, I.; Ibarguren, A.; Kildal, J.; Susperregi, L.; Sierra, B. Human–robot collaboration in industrial applications: Safety, interaction and trust. *Int. J. Adv. Robot. Syst.* **2017**, *14*. [CrossRef]
13. de Gea Fernández, J.; Mronga, D.; Günther, M.; Wirkus, M.; Schröer, M.; Stiene, S.; Kirchner, E.; Bargsten, V.; Bänziger, T.; Teiwes, J.; et al. iMRK: Demonstrator for Intelligent and Intuitive Human-Robot Collaboration in Industrial Manufacturing. *KI-Künstliche Intell.* **2017**, *31*, 203–207. [CrossRef]
14. Raiola, G.; Restrepo, S.S.; Chevalier, P.; Rodriguez-Ayerbe, P.; Lamy, X.; Tliba, S.; Stulp, F. Co-manipulation with a library of virtual guiding fixtures. *Auton. Robot.* **2018**, *42*, 1037–1051. [CrossRef]
15. Munzer, T.; Toussaint, M.; Lopes, M. Efficient behavior learning in human-robot collaboration. *Auton. Robot.* **2018**, *42*, 1103–1115. [CrossRef]
16. Peternel, L.; Tsagarakis, N.; Caldwell, D.; Ajoudani, A. Robot adaptation to human physical fatigue in human–robot co-manipulation. *Autonomous Robots* **2018**, *42*, 1011–1021. [CrossRef]
17. Rozo, L.; Calinon, S.; Caldwell, D.G.; Jimenez, P.; Torras, C. Learning physical collaborative robot behaviors from human demonstrations. *IEEE Trans. Robot.* **2016**, *32*, 513–527. [CrossRef]
18. Mazhar, O.; Ramdani, S.; Navarro, B.; Passama, R.; Cherubini, A. Towards real-time physical human–robot interaction using skeleton information and hand gestures. In Proceedings of the 2018 IEEE/RSJ International Conference on Intelligent Robots and Systems (IROS), Madrid, Spain, 1–5 October 2018; IEEE: Piscataway, NJ, USA, 2018; pp. 1–6.
19. Zhao, R.; Drouot, A.; Ratchev, S. Classification of Contact Forces in Human-Robot Collaborative Manufacturing Environments. *SAE Int. J. Mater. Manuf.* **2018**, *11*, 5–10. [CrossRef]
20. Gaz, C.; Magrini, E.; De Luca, A. A model based residual approach for human–robot collaboration during manual polishing operations. *Mechatronics* **2018**, *55*, 234–247. [CrossRef]
21. Losey, D.P.; McDonald, C.G.; Battaglia, E.; O'Malley, M.K. A review of intent detection, arbitration, and communication aspects of shared control for physical human–robot interaction. *Appl. Mech. Rev.* **2018**, *70*, 010804. [CrossRef]

22. Mohammad, Y.; Xu, Y.; Matsumura, K.; Nishida, T. The H^{3}R explanation corpus human-human and base human–robot interaction dataset. In Proceedings of the 2008 International Conference on Intelligent Sensors, Sensor Networks and Information Processing, Sydney, NSW, Australia, 15–18 December 2008; IEEE: Piscataway, NJ, USA, 2008; pp. 201–206.

23. Jayagopi, D.B.; Sheikhi, S.; Klotz, D.; Wienke, J.; Odobez, J.M.; Wrede, S.; Khalidov, V.; Nguyen, L.; Wrede, B.; Gatica-Perez, D. *The Vernissage Corpus: A Multimodal Human-Robot-Interaction Dataset*; Technical Report; Idiap Research Institute: Martign, Switzerland, 2012.

24. Bastianelli, E.; Castellucci, G.; Croce, D.; Iocchi, L.; Basili, R.; Nardi, D. HuRIC: A Human Robot Interaction Corpus. In Proceedings of the Ninth International Conference on Language Resources and Evaluation, Reykjavik, Iceland, 26–31 May 2014; European Language Resources Association: Paris, France, 2014; pp. 4519–4526.

25. Lemaignan, S.; Kennedy, J.; Baxter, P.; Belpaeme, T. Towards "machine-learnable" child-robot interactions: The PInSoRo dataset. In Proceedings of the IEEE Ro-Man 2016 Workshop on Long-Term Child-Robot Interaction, New York, NY, USA, 31 August 2016.

26. Celiktutan, O.; Skordos, E.; Gunes, H. Multimodal human-human–robot interactions (mhhri) dataset for studying personality and engagement. *IEEE Trans. Affect. Comput.* **2017**. [CrossRef]

27. Yu, K.T.; Bauza, M.; Fazeli, N.; Rodriguez, A. More than a million ways to be pushed. A high-fidelity experimental dataset of planar pushing. In Proceedings of the 2016 IEEE/RSJ International Conference on Intelligent Robots and Systems (IROS), Daejeon, Korea, 9–14 October 2016; IEEE: Piscataway, NJ, USA, 2016; pp. 30–37.

28. De Magistris, G.; Munawar, A.; Pham, T.H.; Inoue, T.; Vinayavekhin, P.; Tachibana, R. Experimental Force-Torque Dataset for Robot Learning of Multi-Shape Insertion. *arXiv* **2018**, arXiv:1807.06749.

29. Huang, Y.; Sun, Y. A Dataset of Daily Interactive Manipulation. *arXiv* **2018**, arXiv:1807.00858.

30. Berndt, D.J.; Clifford, J. Using dynamic time warping to find patterns in time series. In Proceedings of the 3rd International Conference on KDD Workshop, Seattle, WA, USA, 31 July–1 August 1994; AAAI: Menlo Park, CA, USA; Volume 10, pp. 359–370.

31. Bagnall, A.; Bostrom, A.; Large, J.; Lines, J. The great time series classification bake off: An experimental evaluation of recently proposed algorithms. Extended version. *arXiv* **2016**, arXiv:1602.01711.

32. Salvador, S.; Chan, P. Toward accurate dynamic time warping in linear time and space. *Intell. Data Anal.* **2007**, *11*, 561–580. [CrossRef]

33. Shokoohi-Yekta, M.; Hu, B.; Jin, H.; Wang, J.; Keogh, E. Generalizing DTW to the multi-dimensional case requires an adaptive approach. *Data Min. Knowl. Discov.* **2017**, *31*, 1–31. [CrossRef]

34. k-Nearest Neighbors Classifier (Scikit-Learn). Available online: https://scikit-learn.org/stable/modules/generated/sklearn.neighbors.KNeighborsClassifier (accessed on 1 November 2019).

35. Lawrence, N.D. Gaussian process latent variable models for visualization of high dimensional data. In *Advances in Neural Information Processing Systems*; MIT Press: Boston, MA, USA, 2004; pp. 329–336.

36. Tipping, M.E.; Bishop, C.M. Probabilistic principal component analysis. *J. R. Stat. Soc. Ser. B Stat. Methodol.* **1999**, *61*, 611–622. [CrossRef]

37. Su, B.; Ding, X.; Wang, H.; Wu, Y. Discriminative dimensionality reduction for multi-dimensional sequences. *IEEE Trans. Pattern Anal. Mach. Intell.* **2018**, *40*, 77–91. [CrossRef]

38. Villalobos, K.; Diez, B.; Illarramendi, A.; Goñi, A.; Blanco, J.M. I4tsrs: A system to assist a data engineer in time-series dimensionality reduction in industry 4.0 scenarios. In Proceedings of the 27th ACM International Conference on Information and Knowledge Management, Turin, Italy, 22–26 October 2018; ACM: New York, NY, USA, 2018; pp. 1915–1918.

39. Seifert, B.; Korn, K.; Hartmann, S.; Uhl, C. Dynamical Component Analysis (DyCA): Dimensionality reduction for high-dimensional deterministic time-series. In Proceedings of the 2018 IEEE 28th International Workshop on Machine Learning for Signal Processing (MLSP), Aalborg, Denmark, 17–20 September 2018; IEEE: Piscataway, NJ, USA, 2018; pp. 1–6.

40. GPy: Gaussian Process (GP) Framework in Python. Available online: https://sheffieldml.github.io/GPy (accessed on 1 November 2019).

41. Support Vector Machine (Scikit-Learn). Available online: https://scikit-learn.org/stable/modules/generated/sklearn.svm.SVC (accessed on 1 November 2019).
42. Liu, D.C.; Nocedal, J. On the limited memory BFGS method for large scale optimization. *Math. Program.* **1989**, *45*, 503–528. [CrossRef]

Article

Optimized Proportional-Integral-Derivative Controller for Upper Limb Rehabilitation Robot

M. Kamran Joyo [1,*], **Yarooq Raza** [1], **S. Faiz Ahmed** [2], **M. M. Billah** [1], **Kushsairy Kadir** [1,*], **Kanendra Naidu** [1], **Athar Ali** [3] and **Zukhairi Mohd Yusof** [1]

[1]   Universiti Kuala Lumpur, British Malaysian Institute, Selangor 53100, Malaysia
[2]   College of Engineering, American University of Kurdistan, Kurdistan, Iraq
[3]   Department of Industrial Engineering, University of Trento, 38122 Trento, Italy
[*]   Correspondence: muhammad.kamran@s.unikl.my (M.K.J.); kushsairy@unikl.edu.my (K.K.);
     Tel.: +60-11-2379-2551 (M.K.J.); +60-12-3250375 (K.K.)

Received: 10 June 2019; Accepted: 19 June 2019; Published: 25 July 2019

Abstract: This paper proposes a nature inspired, meta-heuristic optimization technique to tune a proportional-integral-derivative (PID) controller for a robotic arm exoskeleton RAX-1. The RAX-1 is a two-degrees-of-freedom (2-DOFs) upper limb rehabilitation robotic system comprising two joints to facilitate shoulder joint movements. The conventional tuning of PID controllers using Ziegler-Nichols produces large overshoots which is not desirable for rehabilitation applications. To address this issue, nature inspired algorithms have recently been proposed to improve the performance of PID controllers. In this study, a 2-DOF PID control system is optimized offline using particle swarm optimization (PSO) and artificial bee colony (ABC). To validate the effectiveness of the proposed ABC-PID method, several simulations were carried out comparing the ABC-PID controller with the PSO-PID and a classical PID controller tuned using the Zeigler-Nichols method. Various investigations, such as determining system performance with respect to maximum overshoot, rise and settling time and using maximum sensitivity function under disturbance, were carried out. The results of the investigations show that the ABC-PID is more robust and outperforms other tuning techniques, and demonstrate the effective response of the proposed technique for a robotic manipulator. Furthermore, the ABC-PID controller is implemented on the hardware setup of RAX-1 and the response during exercise showed minute overshoot with lower rise and settling times compared to PSO and Zeigler-Nichols-based controllers.

Keywords: upper limb rehabilitation robot; particle swam optimization (PSO); artificial bee colony (ABC); Ziegler Nichols; Maximum sensitivity

---

1. Introduction

The life expectancy of people, and hence, the number of older adults, is increasing. Elderly people are most vulnerable to strokes. Stroke is among the main causes of limb disabilities and can be fatal [1,2]. According to statistics [3], more than 10 million people suffer from a stroke annually. Consequently, patients become dependent on others for their basic life activities. Physical therapy under the supervision of physiotherapists can help stroke patients to restore the functionality of their disabled limbs. However, traditional physical therapy involves manpower and is highly expensive as the number of patients increases. Rehabilitation robots have been introduced recently to reduce the burden on physiotherapists and increase the number of exercises performed during a therapy session. These robots can provide a more reliable service as they do not face monotony and fatigue failures due to the repetitive nature of the exercises. One of the most widely used control mechanisms employed in such robots is a proportional–integral–derivative (PID) controller. This mechanism is famous for its simple structure and robust performance in a wide range of operating conditions [4]. However, it is quite difficult to select and determine the PID controller parameters for the system.

The control parameters are tuned to achieve stable closed-loop response of the system and reach desired positions within a certain time. Several approaches have been proposed to optimize the PID controller parameters, such as Zeigler-Nichols, a classical method to tune PID control parameters [5], fuzzy logic controller [6,7] and evolutionary algorithms such as genetic algorithm (GA) [8], and swarm optimization algorithms such as ant colony algorithm (ACO) [9], particle swarm optimization (PSO) [10] and artificial bee colony (ABC) algorithm [11].

The Zeigler-Nichols is a heuristic algorithm based on increasing proportional gain until it reaches the ultimate gain at which output of a control loop system has consistent oscillations. The parameters K_u and oscillation period T_u are used to parametrize PID gains. Using such a controller produces oscillations which introduce large overshoot, higher rise times and lower settling time in the response.

GA is based on the theory of biological operations and optimizes parameters using mutations and crossover operations [12]. ACO is an evolutionary algorithm inspired by the social behavior of ants searching for food using the shortest path. The success rate of ACO is lower than that of PSO [13]. Fuzzy logic is another technique to optimize the parameters of PID. The rehabilitation robot is controlled by PID and the parameters of PID are optimized using fuzzy logic. Every parameter of the PID controller is characterized by different sets of fuzzy rules, and triangular membership functions are used. Experimental results showed that fuzzy PID offers improved and more effective trajectory tracking performance compared to conventional PID controllers [14,15].

PSO algorithm is an evolutionary computation technique inspired by the social behavior of swarms and fish schools. It was first introduced by Eberhart and Kennedy in 1995. Later Eberhart and Shi introduced inertial weights for PSO to provide the global and local exploration balance. PSO has been found to be robust in optimizing nonlinear problems. The PSO technique evaluates the particle position and velocity, which are updated on every iteration with the aim to reach a global best position of the swarm. Every particle in PSO is treated as a volume-less particle in the search space. PSO requires fewer computational resources than GA, which is prone to premature convergence, while its convergence rate is slower than PSO and ABC [16]. PSO is widely used in many engineering applications because of its advancements. Previously, a PSO based optimized PID controller was used to control a multi fingered robotic hand. Comparison of PSO and conventional PID with fuzzy PID is also presented in the work which shows the better results of PSO-PID [17].

ABC is a heuristic technique which is inspired by the intelligent foraging behavior of honeybees. It was proposed by Karaboga in 2005 [11], and is a simple, robust and population based stochastic optimization algorithm [16,18]. The following three categories of bees make the algorithm unique as compared to other swarm algorithms: employed, onlookers and scout bees [19]. In particular, this meta-heuristic algorithm replicates the foraging behavior of honey bees, where a foraging bee evaluates several characteristics of a food source, such as richness of nectar and the complexity of extracting the energy, and communicates the position of the food source to unemployed bees. The communicated information includes the direction and distance to the food source and its profitability; it is regularly updated so that the best food source can be determined. In a recent study, a comparison reported that by using integral square error (ISE) as the objective function to optimize PID, better performance was obtained using PSO than ABC [14]. However, in terms of transient time response that overshoots in the response of system, the rise and settling times of ABC is better than PSO. Although the rise time of PSO is faster than ABC, in rehabilitation, high speed is not recommended, so the slow rise times of ABC eventually offer the benefit of safe movement of the robot and avoid abrupt movements. Also, the ABC algorithm outputs high quality solutions in terms of fitness value with fewer function evaluations in comparison to PSO. It was found that ABC is more robust than PSO optimized controller [20].

This paper presents a comparative analysis of Zeigler-Nichols, ABC and PSO to determine the tuning parameters of a two-degree-of-freedom (2-DOF) PID controller for robotic arm exoskeleton RAX-1. The objective of applying the mentioned optimization algorithms is to establish the optimal control parameters of the 2-DOF PID by minimizing cost, and to meet the prescribed performance criteria. Four different objective functions, i.e., integral square error (ISE), integral time square error

(ITSE), integral absolute error (IAE) and integral time absolute error (ITAE), have been investigated to find the controller with optimal or near optimal load disturbance response subject to robustness and maximum sensitivity constraints. Maximum sensitivity represents the inverse of the minimum distance on the Nyquist plot between critical point and loop transfer function. Such a method for tuning the controller parameters has proven to be effective for robust performance [21]. Furthermore, performance parameters such as overshoot, rise time, settling time and maximum sensitivity are normalized and the least average error (LAE) is evaluated. The optimal solution found for 2-DOF PID for RAX-1 is then implemented with the RAX-1 hardware for trials with three healthy subjects. The rest of the paper is organized as follows. Section 2 explains the controller design, Section 3 presents the simulation results, Section 4 provides the comparative analysis and discussion. Section 5 concludes the paper.

2. Methodology

RAX-1 is an exoskeleton device meant to be used for rehabilitation of upper limb extremities. Operating alongside the human arm, exoskeleton devices are required to produce movements similar to those performed by the upper limb. There are nine DOFs in the upper limb, excluding finger joints. This study focuses on the glenohumeral joint in the shoulder, which is a complex ball-and-socket joint that enables the shoulder to perform movements in three DOFs. These movements are commonly referred to as shoulder extension/flexion, abduction/adduction and medial/lateral rotation, also known as internal/external rotation. Figure 1 represents the three movements that can be performed with the shoulder joint. The ranges of motion for the shoulder joint movements performed by a healthy subject are listed in Table 1 [22]. These movement protocols are then implemented on RAX-1.

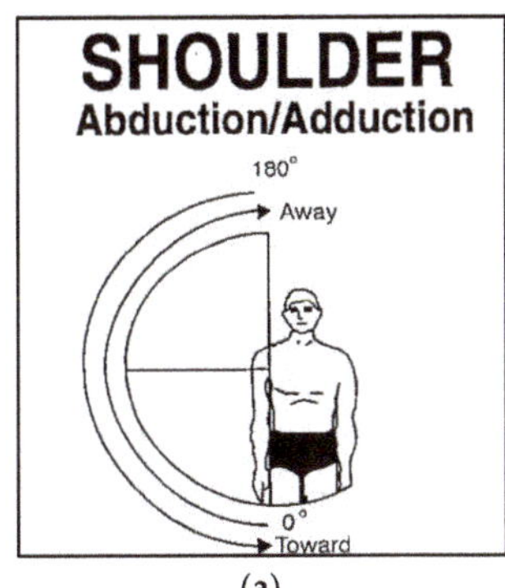

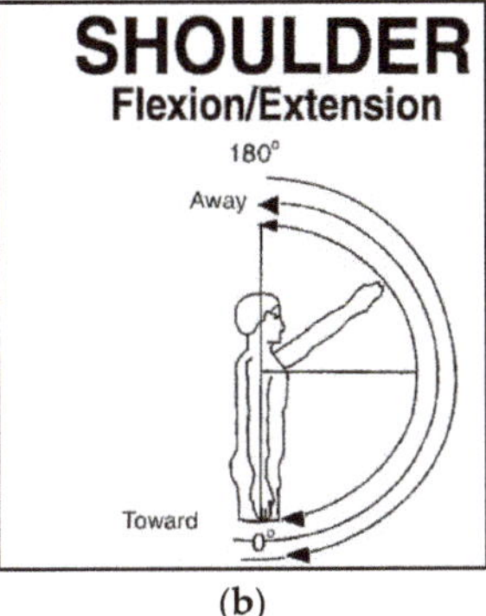

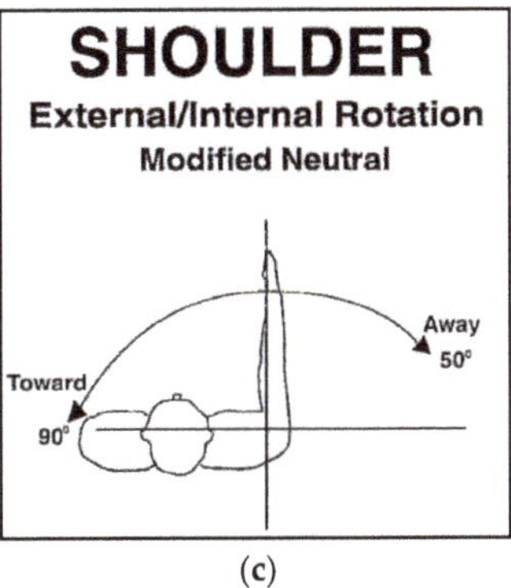

(a)　　　(b)　　　(c)

Figure 1. (**a**) Shoulder abduction and adduction, (**b**) Shoulder Extension/Flexion, (**c**) Shoulder External and Internal rotation [22].

Table 1. Standard ranges of motion of Upper Limb.

Limb	Therapeutic Exercise	ROM of Limb
	Flexion/extension	0°/180°
Shoulder	External/internal rotation	50°/90°
	Abduction/adduction	0°/180°

2.1. System Design

The robotic manipulator in the present study comprises two shoulder joints. The DOF of the manipulator can be calculated by the numbers of links and joints. The 3D model of the robot is illustrated in Figure 2.

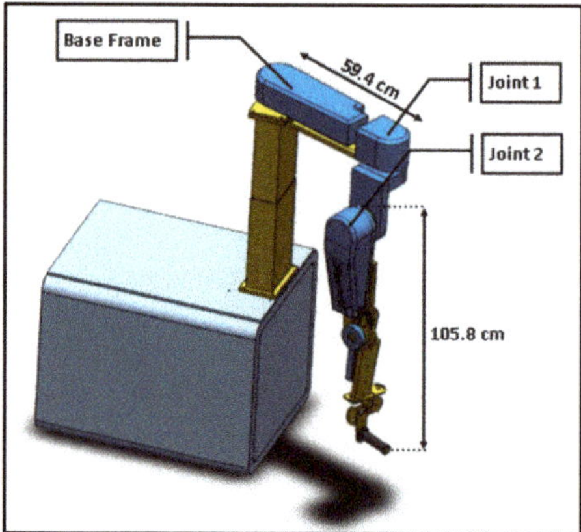

Figure 2. 3D CAD Model of Robot.

A robot in the planner configuration is defined with three parameters (x, y, θ). However, robots are three dimensional in real-world applications; hence, there is a need for six parameters (x, y, z, yaw, pitch, roll) to describe the position and orientation of a robot in space.

Figure 3 represents the direct kinematics of the robotic arm. Every rigid body in a serial chain has a label: Link 1 is the rigid body attached to the shoulder joint 1, Link 2 is the rigid body attached to Link 1 and so on. A joint is present between each link. Hence, Joint 1 attaches Link 1 to Link 0 and Joint 2 attaches Link 2 to Link 1. Frame of reference is numbered according to the respective links they are attached to, e.g. Frame 1 is attached to Link1. Eventually, the aim is to calculate the position of Frame 2 relative to that of Frame 0.

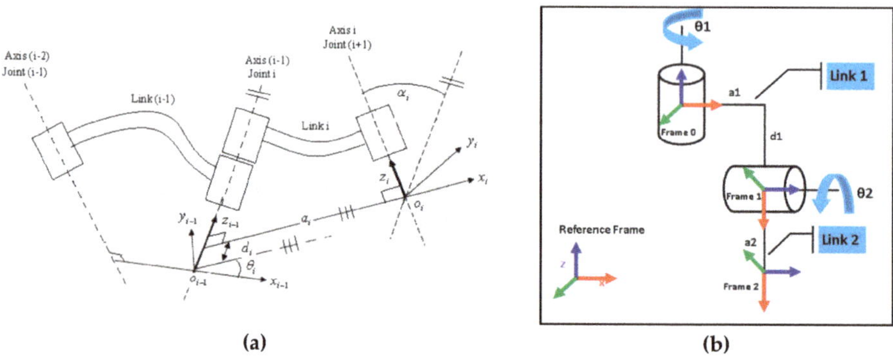

Figure 3. (**a**) DH convention for frame assignment, (**b**) Kinematics of the Robotic Arm.

Figure 3a shows a pair of adjacent links which are link $(i-1)$, and link (i) with their associated joints, joint $(i-1)$, and joint (i). A frame (i) is assigned to link (i) as follows.

1. The z_{i-1} lies along the axis of motion of the ith joint.
2. The x_i axis is normal to the z_{i-1} axis and pointing away from it.

The Denavit-Hartenberg (DH) parameters of a rigid link depends on four geometric parameters $(a_i, \alpha_i, d_i, \theta_i)$ [23]. The four parameters describe any revolute joint as follows:

1. a_i (Link length) is a distance measured along the x_i axis from the point of intersection of x_i axis with z_{i-1} axis to the origin of frame (i).

2. α_i (Link Twist) is the angle between the joint axes z_{i-1} and z_i axes measured about x_i axis in the right-hand orientation.

3. d_i (offset) is the distance measured along z_{i-1} axis from the origin of frame $(i-1)$ to the intersection of x_i axis with z_{i-1} axis.

4. θ_i (Joint angle) is the angle between x_{i-1} and x_i axes measured about the z_{i-1} axis in the right-hand sense.

The 2-DOF upper limb robotic manipulator can be calculated from Figure 3b. The transformation matrix from Frame 0 to the end-effector can be defined as:

$$
T_0^e = \begin{bmatrix}
c_{12} & c_1 s_2 & -s_1 & a_2 c_{12} + a_1 c_1 \\
s_1 c_2 & -s_{12} & c_1 & a_2 s_1 c_2 + a_1 s_1 \\
-s_2 & -c_2 & 0 & a_2 s_2 + d_1 \\
0 & 0 & 0 & 1
\end{bmatrix}
$$

where $a_1 = 59.4$ cm, $a_2 = 105.8$ cm and $d_1 = 20$ cm. The transformations can be used to determine kinematic measurements of the joints. For any joint angle, the position of the end effector can be derived from the transformation matrix.

2.2. Dynamic Model

In this study, Euler-Lagrangian approach was applied to calculate the dynamics of the robot manipulator. This approach uses the joint velocities and position to determine the kinetic and potential energies of a system. It generalizes Newtonian mechanics for systems that are subject to a specific class of constraints. These constraints are often expressed in terms of the position or variables describing the system in question.

The Lagrangian equation of motion defined in (1) is written as:

$$
\frac{d}{dt}\left(\frac{\partial L}{\partial \dot{q}_j}\right) - \frac{\partial L}{\partial q_j} = \tau_j \tag{1}
$$

where τ_j denotes the required torque, $L = K - P$ (Kinetic and potential energies) is the Lagrangian and qj is the generalized coordinate of the jth joint of robot.

The inertial matrix $D(q)$ can be determined as follows:

$$
D(q) = m_1 J_{vc1}^T J_{vc1} + m_2 J_{vc2}^T J_{vc2} + \begin{bmatrix} I_1 + I_2 & 0 \\ 0 & I_2 \end{bmatrix}, \tag{2}
$$

where $m_1 = 1.5$ kg, $m_2 = 0.5$ kg. Simplifying (2), one can obtain the following (3).

$$
\begin{aligned}
d_{11} &= m_1 a_{c1}^2 + 2m_2 a_1 a_{c2} c_2 - 2a_{c2} d_1 m_2 s_2 + a_1^2 m_2 + a_{c2}^2 m_2 + d_1^2 m_2 + I_1 + I_2 \\
d_{12} &= m_2 a_1 a_{c2} c_2 - m_2 d_1 a_{c2} s_2 + m_2 a_{c2}^2 \\
d_{21} &= m_2 a_1 a_{c2} c_2 - m_2 d_1 a_{c2} s_2 + m_2 a_{c2}^2 \\
d_{22} &= m_2 a_{c2}^2 + I_2
\end{aligned} \tag{3}
$$

The correction term Christoffel symbols ensures that when the derivatives of the vector field lying in a tangent plane of the configuration manifold are computed, they stay in the same tangent space. The Christoffel symbols in (4) are defined as

$$\left. \begin{aligned}
c_{111} &= \frac{\partial d_{11}}{\partial q_1} = 0 \\
c_{121} = c_{121} &= \tfrac{1}{2}\cdot\frac{\partial d_{11}}{\partial q_2} = -m_2 a_1 a_{c2} s_2 - m_2 d_1 a_{c2} c_2 = h \\
c_{221} &= \frac{\partial d_{12}}{\partial q_2} - \tfrac{1}{2}\cdot\frac{\partial d_{22}}{\partial q_1} = h \\
c_{112} &= \frac{\partial d_{21}}{\partial q_1} - \tfrac{1}{2}\cdot\frac{\partial d_{11}}{\partial q_2} = -h \\
c_{122} = c_{212} &= \tfrac{1}{2}\cdot\frac{\partial d_{22}}{\partial q_1} = 0 \\
c_{222} &= \tfrac{1}{2}\cdot\frac{\partial d_{22}}{\partial q_2} = 0
\end{aligned} \right\} \tag{4}$$

The potential energy of each joint P_i, is the product of the mass of that link m_i, position vector to the centre of mass r_{ci} and acceleration due to gravity g:

$$P_i = m_i g r_{ci};\ i = 1,2 \tag{5}$$

The term ϕ_k is a function of generalized coordinates that does not depend on their derivatives. It is given by the partial derivative of potential energy of the system with respect to the generalized coordinates as follows:

$$\phi_k = \frac{\partial P}{\partial q_k} \tag{6}$$

Finally, the dynamic equations of the system after substituting various quantities and omitting zero can be expressed as

$$\left. \begin{aligned}
d_{11}\ddot{q}_1 + d_{12}\ddot{q}_2 + c_{121}\dot{q}_1\dot{q}_2 + c_{211}\dot{q}_2\dot{q}_1 + c_{211}\dot{q}_2^{\,2} + \phi_1 = \tau_1 \\
d_{21}\ddot{q}_1 + d_{22}\ddot{q}_2 + c_{112}\dot{q}_1^{\,2} + \phi_2 = \tau_2
\end{aligned} \right\} \tag{7}$$

which, in general can be written in matrix form as:

$$D(q)\ddot{q} + C(q,\dot{q})\dot{q} + g(q) = \tau \tag{8}$$

Here, $D(q)$ denotes the inertia matrix of the system and $C(q,\dot{q})$ gives the Christoffel symbols and $g(q)$ is actually ϕ_k which is determined by taking partial derivative of potential energy with generalized coordinates. τ is a 2×1 matrix representing the generalized active forces.

2.3. Linearized Model

The linearized state space model for robot exoskeleton (RAX-1) is expressed as follows.

$$A = \begin{bmatrix} 0 & -5.5510e-13 & 0 & 0 \\ 1 & 0 & 0 & 0 \\ 0 & 0 & 0 & -25.8600 \\ 0 & 0 & 1 & 0 \end{bmatrix}, B = \begin{bmatrix} 1 & 0 \\ 0 & 0 \\ 0 & 1 \\ 0 & 0 \end{bmatrix}$$
$$C = \begin{bmatrix} 0 & 31.66 & 0 & 0 \\ 0 & 0 & 0 & 6.3760 \end{bmatrix}, D = \begin{bmatrix} 0 & 0 \\ 0 & 0 \end{bmatrix} \tag{9}$$

2.4. Motor Model

The robot manipulator requires actuators to provide the desired amount of torque at the joints. The actuators convert electrical energy into rotational mechanical energy. DC motors are widely used in robotics as actuators due to their high torque, speed controllability and portability [24]. The internal model of DC motor is illustrated in Figure 4 and can be expressed as follows.

$$\frac{\theta(s)}{E_a(s)} = \frac{K_\tau}{(L_a s + R_a)(Js + B_m) + K_b K_\tau} \approx \frac{K_\tau}{R_a(Js + B_m) + K_b K_\tau} \tag{10}$$

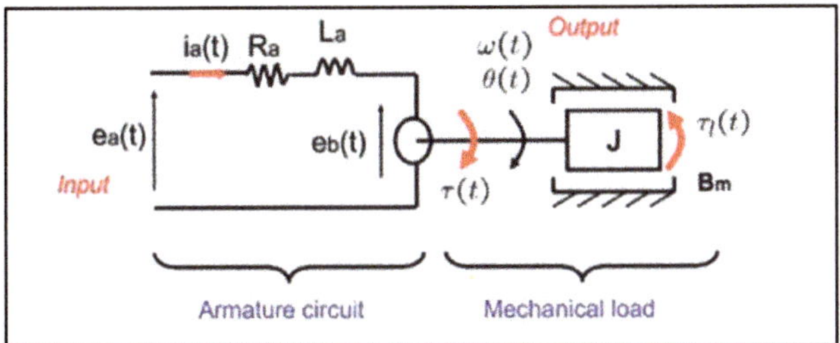

Figure 4. Internal model of DC motor.

Here, 'J' is the motor inertia, 'B_m' motor damper, 'K_τ' is the motor constant, 'K_b' is the proportionality constant between angular velocity of the motor and back emf, whereas 'L_a' and 'R_a' inductance and resistance of the armature. If $L_a \ll R_a$ then, an approximated transfer function of motor is obtained by setting $L_a = 0$. This converts motor model from a second order system to 1st order system. DC motors with harmonic gears are used in this system. Table 2 lists the motor parameters.

Table 2. Motor Parameters.

Parameters	Joint 1	Joint 2
K_τ (mN-m/A)	70.5	70.5
L_a (mH)	0.264	0.264
R_a (Ohm)	0.343	0.343
K_b (V-s/rad)	0.023	0.023
J (g·cm^2)	306×10^{-6}	306×10^{-6}
B_m (N.sec/m)	0.03	0.03
Gear Ratio (Nm/Nl)	1/160	1/150

3. The Exoskeleton Platform

In this section, the hardware design for the upper limb extremity rehabilitation is described. The platform consists of the design and manufacturing of the mechanical structure, actuators and sensors with hardware implementation of the control algorithm.

3.1. Mechanical Structure

The 2-DOF mechanical platform is built using aluminum grade 6061 alloy and weighs approximately 15 kg. The structure is specifically designed to focus on specific exercises for rehabilitation of the shoulder joint (Figure 5).

The upper extremity exoskeleton device consists mainly of a support frame, height adjustment mechanism and shoulder actuation mechanism. The support frame is attached to wheels enabling the platform to be remote. The manually controlled height regulation mechanism is attached to the support frame. The actuation mechanism is attached to the height regulation mechanism; it adjusts the mechanical frame fixed to human arm to match requirements to the subject height. The human arm is fixed to the exoskeleton with soft wraps both on the forearm and bicep. Other relevant adjustments are made during gait training. Joints at the wrist and elbow are passive and their orientations are fixed.

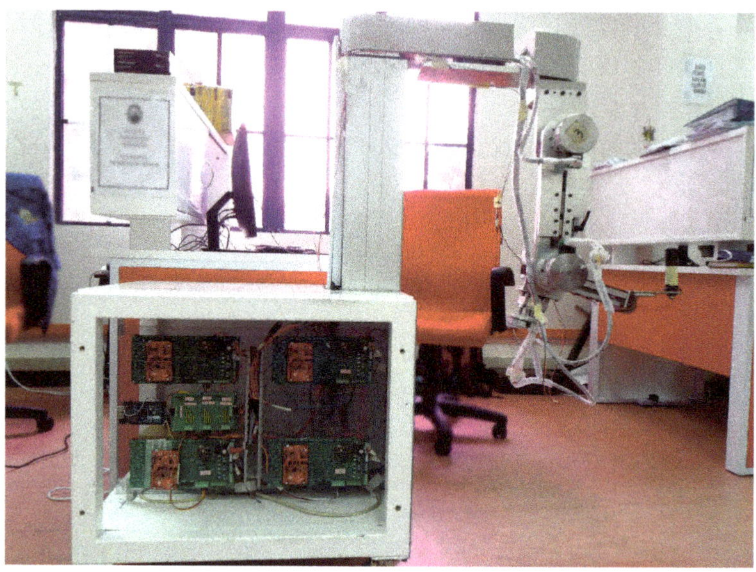

Figure 5. RAX-1 Mechanical design.

3.2. Actuators and Sensors

The two-revolute joint system is equipped with the MAXON EC-90 motor (Maxon Motor, Sachseln, Switzerland) attached to the gear. A brushless flat motor has the maximum angular velocity of 2590 rpm and power rating of 90W which can generate maximum torque of 444 mNm. It is equipped with three Hall effect sensors to measure current and velocity and an internal encoder generating 2040 pulses per rotation. The reducers (harmonic drives) combined with flat motor have speed ratio of 160:1 for the first shoulder joint responsible for the internal/ external movement, while the ratio is 150:1 for the other joint responsible for the extension/ flexion movement. The detailed specification of other motor parameters and gear ratio is provided in Table 2. A force sensor is attached to the wrist handle of the robot, from which external disturbance exerted by the subject is measured while the exoskeleton is working in passive mode. The electrical setup of the system involves an ESCON 50/5 module which acts like a closed-loop speed or current controller for the motor.

3.3. Control Implementation

In the control system, a host computer is used as a graphical user interface. The software is written in visual C#; it allows the user to select one of the possible working modes of the exoskeleton (passive, active or semi-active). The software enables uploading and downloading subject's history over a cloud specifically designed for this device. The user can also define the range of motion at which the exoskeleton should work for several repetitions.

A master/slave network is designed to interconnect the user with the system, where each joint in the system is a slave. CC3200 from Texas Instrument is used as a peripheral device that controls the angular movement. The master is instructed by the user to select an exercise and the number repetitions, as well as set an angular movement via an interface. This information is then communicated to the slaves for further implementation of the task [25] as depicted in Figure 6a,b. The overall system block diagram is illustrated in Figure 6c.

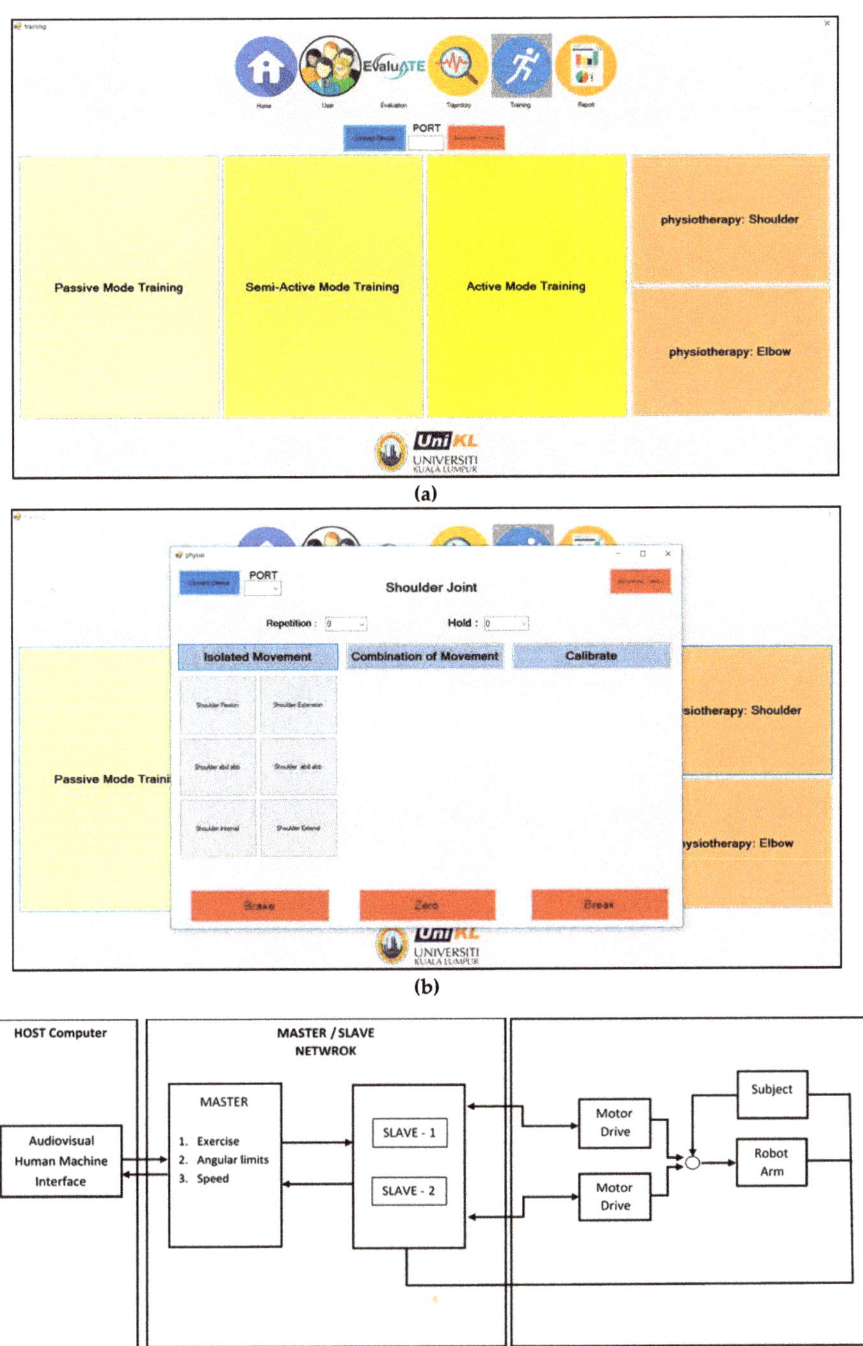

Figure 6. (**a**) HMI displaying rehabilitation training modes, (**b**) HMI displaying selection for shoulder exercise, (**c**) System Block Diagram.

Figure 7 represents the controller card used for RAX-1; it consists of master-slave network and drivers used for motors.

Figure 7. Controller Board.

4. The Control Algorithm

4.1. Proportional Integrator Derivative (PID) Controller

One of the most widely used controllers is the PID controller, which has a very simple structure and is robust in under wide range of operating conditions. The PID error signal, which is the difference between setpoint and measured variable, is calculated and fed back into the system continuously to change the proportional, integral and derivative gains accordingly [26].

PID is one of the most used control algorithms in industrial control systems. The response of a system is categorized by the rise time, overshoot, settling time and steady state error. In this study, a 2-DOF PID controller is used, and is represented mathematically according to Equation (11). The controller is capable of rejecting disturbances without significant increase of overshoot in setpoint tracking. It includes setpoint weighting on the proportional and derivative terms. A typical 2-DOF PID is composed of feed-forward and feedback compensators. The feed-forward compensator consists of a PD component while the feedback compensator includes PID component as shown in Figure 8.

$$u = K_p(b.r - y) + K_i(r - y)s^{-1} + K_d \frac{N}{1 + Ns^{-1}}(c.r - y) \tag{11}$$

where u denotes the input given to the plant or system, while K_p, K_d, K_i, denotes the proportional, derivative and integral gains. In this study, 2-DOF PID in MATLAB is used, which has three more parameters to tune; b, c and N, where; b and c denote the setpoint weights, while N denotes a filter coefficient.

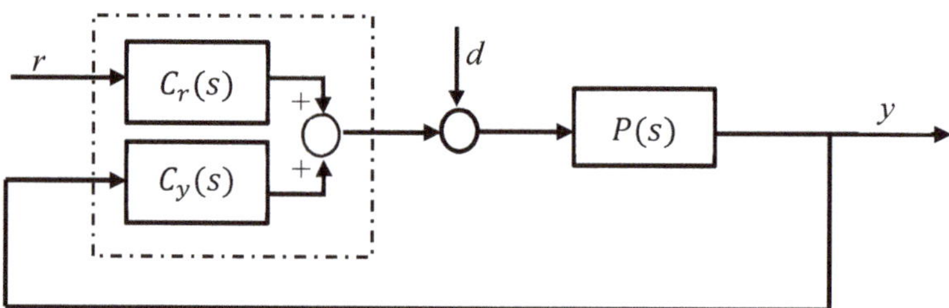

Figure 8. 2-DOF PID block diagram.

4.2. Particle Swam Optimization (PSO)

PSO is an optimization technique inspired by the social behavior found in nature, such as flocking of birds and fish schooling. In the PSO search space, each solution acts like a flying bird in the search of food, known as a particle. PSO works based on the social behavior of particles in a swarm. This algorithm locates and finds the global best solution by adjusting an objective function. At the end of the process, best solution based on objective function is found for 2-dof pid controller parameters.

First, PSO chooses random solutions to initialize the population. Then, it updates its performance to obtain optimum value. Every particle is characterized by its position and velocity in the swarm. The velocity of a moving particle depends upon the change in position or direction [27]. Each particle updates its new position based on two components, p_{best} and g_{best}, where p_{best} is the best position attained by a particle, while g_{best} is the global best position of the entire swarm. The velocity and position of the particle can be expressed according to Equations (10) and (11), respectively.

$$v_{ij}(t+1) = wv_{ij}(t) + r_1 c_1 \left(p_{ij}(t) - x_{ij}(t) \right) + r_2 c_2 \left(g_j(t) - x_{ij}(t) \right) \tag{12}$$

$$x_{ij}(t+1) = x_{ij}(t) + v_{ij}(t+1) \tag{13}$$

where; $v_{ij}(t)$ denotes the particle velocity, $x_{ij}(t)$ denotes the particle position, $v_{ij}(t+1)$ denotes the particle updated velocity, $x_{ij}(t+1)$ denotes the particle updated position; $p_{ij}(t)$ denotes the particle best position; $g_j(t)$ denotes the global best position of the swarm; w, $(w = w_{max} - w_{min})$ denotes the inertia term; r_1 and r_2 are two uniformly distributed random numbers ranging from 0 to 1 and c_1 and c_2 are the acceleration coefficients. The important steps of the PSO are summarized in Figure 9.

According to the PSO algorithm, the swarm size, position, velocity and the constants w, c_1 and c_2 are initialized first. Then, the fitness value of each particle is calculated, P_{best} and g_{best} are defined and the position and velocity of each particle are updated. The algorithm stops when the stopping criterion is met. The optimal solution is chosen according to the latest g_{best}. The PSO parameters and their values used in our study are provided in Table 3.

Table 3. PSO Parameters.

Parameter	Value
Number of particles	20
Number of iterations	150
w_{min}	0.4
w_{max}	0.9
c_1	2
c_2	2

4.3. Artificial Bee Colony (ABC)

ABC is an optimization technique based on the smart behavior of honey bees [16], which are famous for their intelligence and ability to perform complex tasks such as collecting nectar and building nests with a high degree precision [28]. Information about the quality of a food source is communicated within the colony by a particular dance language. The precision of the foraging range of honeybees allows an efficient exploitation of food sources and concentration of foraging on the best patches. There are two main concepts that describe swarm intelligence, namely self-organization and labor division [29]. The self-organizing behavior represents the complex collective behavior that rises from the local interaction between the agents showing a simple self-directed behavior. The mechanism of labor division assigns specific tasks to the agents performing simultaneous activities, which results in a more efficient and time-saving performance.

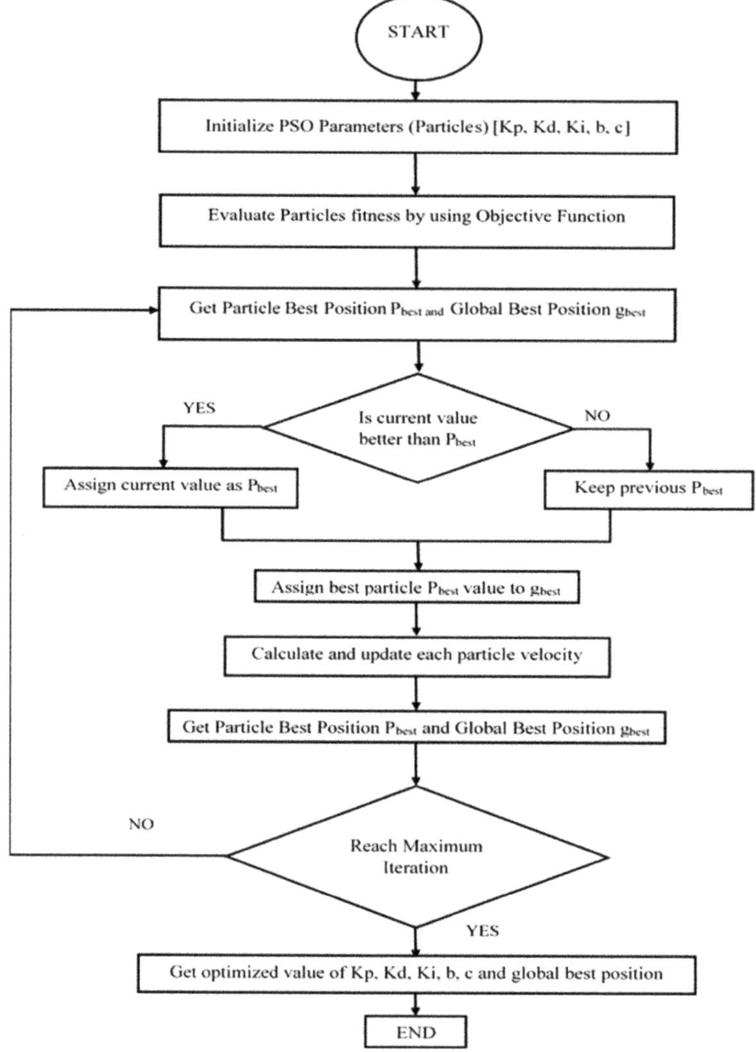

Figure 9. PSO Optimization Flow.

There are three categories of artificial bees in ABC algorithm: employed, scout and onlooker bees. The employed bees go to the food source that has already been visited by them, while unemployed bees look for further sources of food. The number of employed bees is equal to that of food sources. Searching for new sources is performed by the scout bees, whereas the onlooker bees wait for the information about the discovered food sources provided by employed bees via their waggle dance. If the position of a food source does not improve within a number of attempts known as limits, then the employed bees become scout bees. In this manner, the exploitation process is performed by employed and onlooker bees, whereas the scout bees explore the existing solutions. The details of different ABC phases are described in the following subsections.

a. Initialization Phase

Random initialization of the locations of food sources is performed according to the following equation:

$$x_{ij} = x_j^{max} + rand(0,1)\left(x_j^{max} - x_j^{min}\right); \; i = 1,\dots,SN, \; j = 1,\dots,D \tag{14}$$

where SN is the number of food sources taken as half of the bee colony, D is the dimension of the problem, x_{ij} denotes the ith employed bee on j^{th} dimension, wjile x_j^{max} and x_j^{min} denotes its upper and lower bounds.

b. Employed Bee Phase

Every other employed bee is allocated with the food source for further exploitation. Equation (11) represents the process of generating a food source.

$$v_{ij} = x_{ij} + \phi\left(x_{ij} - x_{kj}\right) \tag{15}$$

where $\phi = a \times rand(0,1)$ is a random variable denoting the acceleration coefficients ranging in the interval $[-1,1]$ and v_{ij} is the new solution or a food source. The fitness fit_i of the new food source is calculated according to the following equation:

$$fit_i = \begin{cases} \frac{1}{1+f_i}, & f_i \geq 0 \\ 1 + abs(f_i), & f_i < 0 \end{cases} \tag{16}$$

where, f_i is the objective function of each food source. A selection is made between the original and new food sources to choose the better one by keeping the fitness value in accordance.

c. Probabilistic Selection Phase

An onlooker bee selects a food source with a certain probability calculates as

$$p_i = \frac{fit_i}{\sum_{j=1}^{N} fit_j} \tag{17}$$

where p_i denotes the probability of selecting the ith solution.

d. Onlooker Bee Phase

The employed bees share the information about food sources with the onlooker bees, who select a food source to exploit better solutions according to its selection probability. The fitness values of each exploited food sources is calculated. A greedy selection between original and new food sources is made similar to the employed bees phase.

e. Scout Bee Phase

If a food source does not yield better results within the limits L, where $L = 0.6 \times$ (*Number of optimization parameters*) $\times$ (*colony size*), then this food source is abandoned and the bee associated with it becomes a scout bee. In this case, a new source of food is randomly generated according to Equation (15). All phases will continue until the termination criterion is met. The output is the best food source solution. Figure 10 shows the flowchart of the ABC algorithm which depicts the process of 2-dof pid controller optimization.

According to the ABC algorithm, the colony size, position and the constants L and a are initialized first. Then, the fitness value is calculated, and the best food source is defined. The algorithm stops when the stopping criterion is met. The optimal solution is chosen according to the latest g_{best}. The ABC parameters and their values used in our study are provided in Table 4.

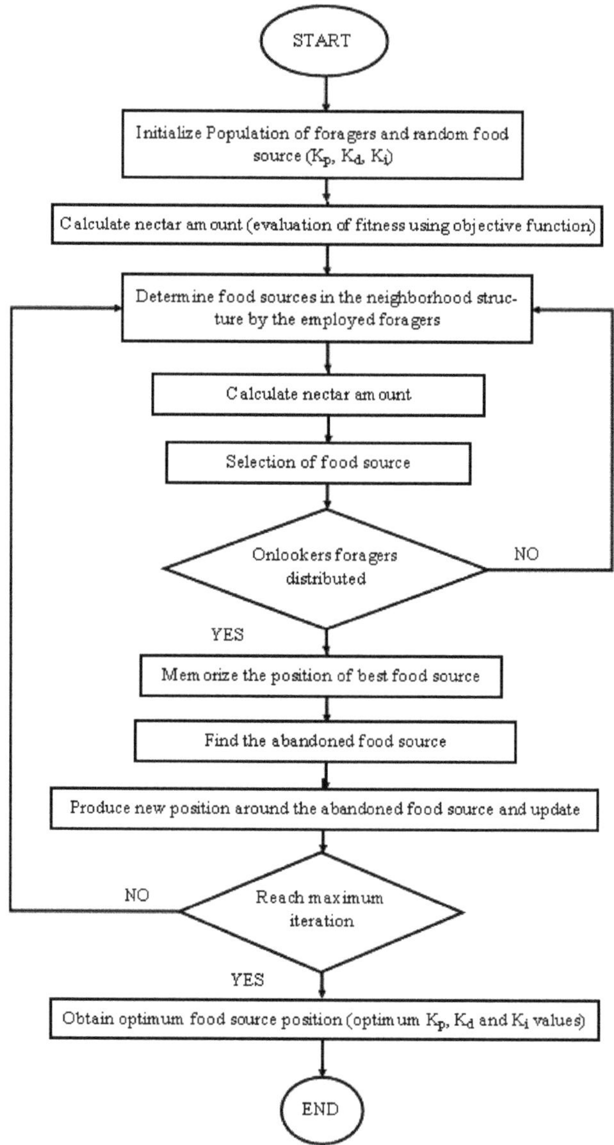

Figure 10. ABC Optimization Flow.

Table 4. ABC Parameters.

Parameter	Value
Colony Size	20
Number of iterations	150
L (Abandonment Limit)	72
a (Acceleration coefficient)	1

5. Results and Discussions

Based on the simulation, the system is divided into two linearized sub systems representing Joint 1 and Joint 2 [30]. Independent controls for both joints with saturation limits are implemented. A step input is given to the model of the robot and the response is observed. A schematic of the proposed controller's tuning method based on PSO and ABC is shown in Figure 11.

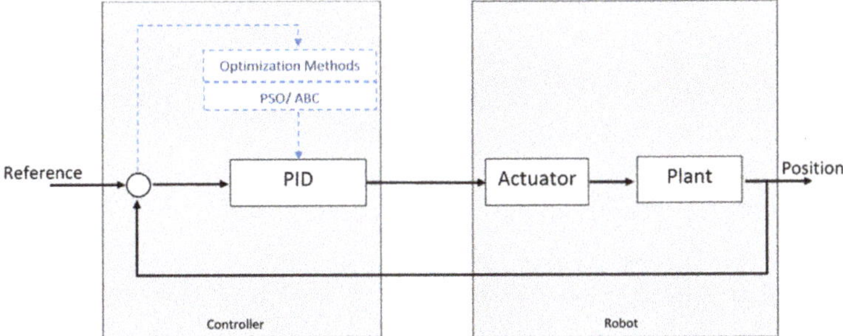

Figure 11. General Structure of the proposed control System Diagram.

For this research, a comparative study was carried out with four different cost functions to gauge the appropriate objective function for this study. A well-chosen objective function leads to better performance of the system and meets control design expectations. These objective functions include the integral squared error (ISE), integral time squared error (ITSE), integral absolute error (IAE) and integral time absolute error (ITAE), and are defined as follows:

$$ISE = \int e^2 dt \tag{18}$$

$$IAE = \int |e| dt \tag{19}$$

$$ITSE = \int t.e^2 dt \tag{20}$$

$$ITAE = \int t.|e| dt \tag{21}$$

5.1. Robustness Consideration

Focus on the tuning procedure requires some robustness considerations in the design. This is achieved by using the maximum sensitivity function as a measure of robustness and is given by

$$M_s = \max_{\omega} |S(j\omega)| = \max_{\omega} \frac{1}{|1 - C_y(j\omega)P(j\omega)|} \tag{22}$$

where $|S(j\omega)| \le M_s$. The sensitivity function shows the effect of feedback on the output. Disturbances are attenuated if $|S(j\omega)| < 1$ and are amplified if $|S(j\omega)| > 1$. The robustness of the closed loop increases with the decrease in M_s. The values for M_s ranging from 1.2 to 2.0 provides reasonable robustness [27].

5.2. Simulation Setup

Before implementing ABC-PID [11] on hardware, simulations are carried out to validate and verify the control algorithm. In this section, a comparative analysis of the PID controller optimized with ABC, PSO, and conventional tuning method based on Zeigler-Nichols is presented. The parametric bounds are provided in Table 5. The Zeigler-Nichols PID parameters and PID parameters tuned using ABC and PSO used for simulation are provided in Tables 6 and 7. The simulations presented here are for internal/external rotation, extension/flexion and abduction/adduction of the shoulder joint.

Table 5. Parametric Constraints used for PSO and ABC.

Parameters	Joint 1	Joint 2
K_p	$0.01 < K_p < 10$	$0.01 < K_p < 20$
K_i	$0.01 < K_i < 10$	$0.01 < K_i < 20$
K_d	$0.01 < K_d < 10$	$0.01 < K_d < 20$
b	$0.01 < b < 1$	$0.01 < b < 1$
c	$0.01 < c < 1$	$0.01 < c < 1$
N	100	200

Table 6. Optimized PID Parameters for Joint 1.

Controller	Objective Function	K_p	K_i	K_d	b	c
ABC-PID	ISE	7.271	9.7748	10	0.9533	0.9826
	IAE	10	9.9485	5.8094	0.9521	0.9198
	ITSE	7.003	9.9345	10	0.9401	0.9993
	ITAE	7.9219	10	2.339	0.8137	0.6124
PSO-PID	ISE	6.8358	10	10	1	1
	IAE	9.9531	8.8488	6.0963	1	0.9535
	ITSE	10	10	9.9951	1	1
	ITAE	10	10	3.3444	1	1
ZN-PID	-	2.7	2.5714	0.7087	1	1

Table 7. Optimized PID Parameters for Joint 2.

Controller	Objective Function	K_p	K_i	K_d	b	c
ABC-PID	ISE	8.7359	10	0.7251	1	0.9717
	IAE	7.229	10	0.5594	0.9999	0.7415
	ITSE	10	10	0.6289	1	0.6926
	ITAE	4.8276	10	0.5214	1	0.8247
PSO-PID	ISE	11.011	20	1.1377	1	1
	IAE	6.1177	20	0.5864	1	0.9625
	ITSE	4.0505	20	0.4883	0.1	0.8429
	ITAE	4.7981	20	0.3883	0.1	0.1
ZN-PID	-	1.92	2.7429	0.3360	1	1

To evaluate the fitness of the objective function, normalization of the objective function is carried out, which scales objective function within a specified range. The following normalization function is used [31].

$$fit'_i = \frac{fit_i - min(fit_{overall}) \times \delta}{max(fit_{overall}) - min(fit_{overall}) \times \delta} \tag{23}$$

where fit_i represents the objective to be normalized and $fit_{overall}$ represents the overall fitness. δ is kept 0.99 to avoid any zeros during normalization. Furthermore, normalized average has been evaluated to determine the significance of the objective function.

Table 8 shows that IAE and ITAE used as objective functions produce minimal overshoots and suitable rise and settling times for both PSO-PID and ABC-PID. When using ISE and ITSE, the rise time is small, while larger overshoots are detected. However, it is obvious that the performance parameters of ABC-PID for Joints 1 and 2 are much better in terms of the six performance parameters mentioned above compared to that of PSO-PID or Zeigler-Nichols.

Table 8. Performance Controller Parameters for all Objective Function.

| | Integral Absolute Error | | | | | | | | | | | | | |
| | Joint 1 | | | | | | | Joint 2 | | | | | | |
Controller	Best Cost	O.S (%)	R.T (s)	S.T (s)	Disturbance O.S (%)	Ms	Normalized Average	Best Cost	O.S (%)	R.T (s)	S.T (s)	Disturbance O.S (%)	Ms	Normalized Average
PSO-PID	0.219	3.1780	0.0641	0.3	6.6	1.24	0.2326	0.0809	22.0742	0.0159	0.1452	11.5	1.62	0.2425
ABC-PID	0.16375	0	0.0750	0.15	6.5	1.23	0.1643	0.1304	4.4458	0.0211	0.1123	11.5	1.62	0.1541
	Integral Squared Error													
PSO- PID	0.0292	6.8484	0.0349	0.9244	6.5	1.37	0.4346	0.0192	46.2938	0.0093	0.1243	6.9	2.0	0.3583
ABC-PID	0.0283	4.9856	0.0360	0.1058	6.4	1.37	0.2476	0.0224	29.9438	0.0131	0.1115	9.6	1.80	0.1983
	Integral Time Squared Error													
PSO- PID	0.1065	7.7363	0.0345	0.254	5.5	1.38	0.3771	0.2021	3.0304	0.5020	0.9666	15.4	1.50	0.3637
ABC-PID	0.1457	6.5768	0.0350	0.1217	6.4	1.37	0.2762	0.0936	5.9488	0.0192	0.1082	8.7	1.80	0.2863
	Integral Time Absolute Error													
PSO- PID	1.8319	17.2245	0.0958	0.9244	7.7	1.15	0.6496	1.0296	0	0.4208	0.6515	15.1	1.43	0.5296
ABC-PID	1.7792	1.8361	0.2748	0.4130	10.2	1.11	0.5541	1.5827	5.1007	0.0216	0.0954	15.6	1.53	0.3708
ZN-PID	-	51.4168	0.2819	4.5369	51.1	1.91	-	-	10.2841	0.0284	1.4003	37.1	1.3168	-

The results plotted in Figure 12 show the response for Joint 1 for PID controller tuned using the Zeigler-Nichols and optimal parameters found with PSO and ABC for all four-objective functions. The figure shows that the ZN response is quicker with a higher rise time, higher settling time and larger overshoot. The results also show a very low percentage of overshoot in the response with a very low steady state error for all four objective functions. However, the optimal response of the system is found while using IAE as objective function.

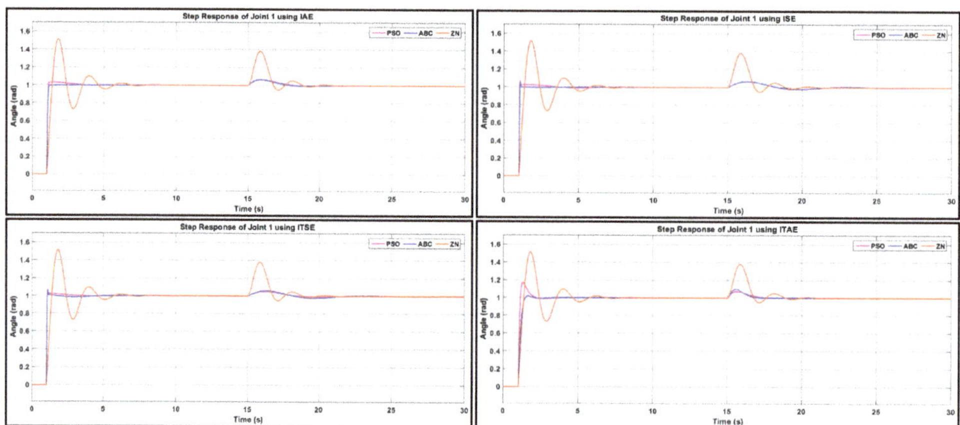

Figure 12. Step response of Joint 1 using ISE, IAE, ITSE and ITAE.

Figure 13 shows the step response for Joint 2. It is evident from the figure that PID tuned Zeigler-Nichols produces large overshoot than the PID optimized with ABC or PSO. However, the overshoot of ABC is 4.4% which is smaller as compared to 22% overshoot of PSO when using IAE. It was found that PID-ABC has the minimum average of the normalized objective function for Joint 1 and Joint 2. Hence, the ABC-optimized PID controller is found to be robust and stable for practical implementation on a robotic arm manipulator. The performance and robustness obtained by calculating the system percent overshoot, rise time, settling time, cost and sensitivity is presented in Table 8.

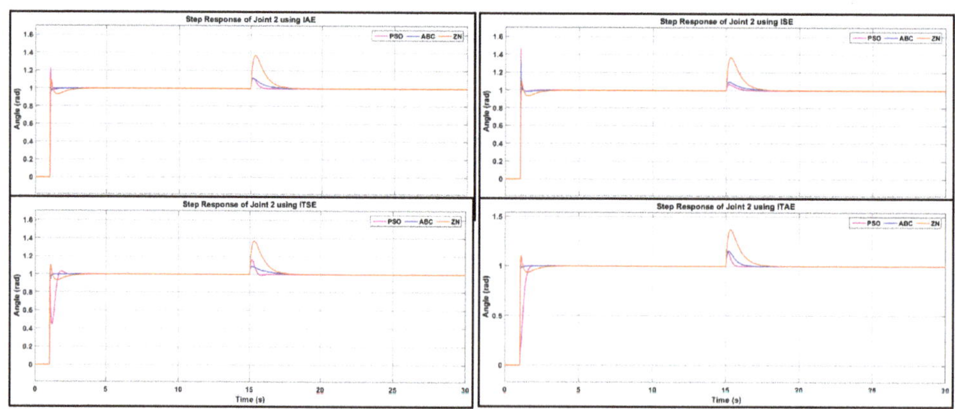

Figure 13. Step response of Joint 2 using ISE, IAE, ITSE and ITAE.

Furthermore, an investigation on closed-loop stability analysis is performed by Nyquist plot of the controllers tuned with ABC using IAE for the given systems. The plot for Joint 1 and Joint 2 are shown in Figure 14. It indicates that the both systems are asymptotically stable as no poles lie in the right half of the plane.

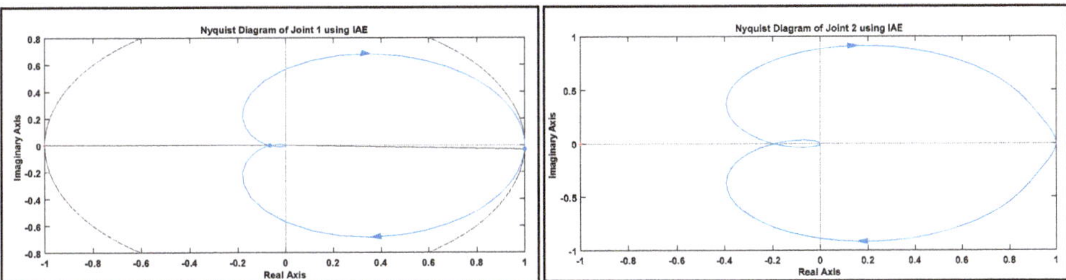

Figure 14. Nyquist plot for closed loop system of joint 1 and joint 2.

5.3. Experimental Evaluation for RAX-1

A preliminary experiment was conducted to test the efficacy and performance of the exoskeleton controlled using the optimized PID controller. Three healthy male subjects participated in the experiment and their properties are listed in Table 9. Subjects were bound with soft wraps and their preliminary tests were performed before and after the exercise. The tests included measuring the body temperature, heart rate and oxygen levels. The evaluation procedure was approved by the Ethics committee of the University of Kuala Lumpur (UniKL), Kuala Lumpur, Malaysia. All subjects were volunteers who signed consent forms before the experiment. The procedure was completed in the presence of physiotherapists from the Royal College of Medicine Perak, Malaysia. The mechanical structure used for this experiment was RAX-1 illustrated in Figure 5.

Table 9. Subject properties.

Parameters	Subject 1	Subject 2	Subject 3
age	32	31	28
body mass (kg)	74	70	64
Height (foot)	5.74	5.91	5.68

In this experiment, the upper extremity exoskeleton, which is a robot in charge of the rehabilitation protocol, provides passive assistance for the subjects. The two actuators follow the specific trajectories of a pre-defined range of motions. The followed trajectory and joint angles can be measured from the encoders. Figure 15 shows a subject performing the shoulder exercises with the assistance of exoskeleton device. The angular rotation of the joint can be set via the interface, and its current position can be observed on the screen.

A ramp/ sinusoidal repetitive input is provided to the shoulder joint to perform three different exercises, namely shoulder extension/flexion, internal/external rotation and abduction/adduction (Table 1).

For this experiment, a combined shoulder external/internal rotation was performed by the first subject; the upper bound movement during external rotation was limited to 90° and that during internal rotation was limited to 0°. Figure 16a illustrates the motion tracking of the robotic arm where a minute overshoot was observed in reaching the target reference. The graph also shows the error representation between the reference and actual response and speed of the motor in terms of rpm. During the experiment, each subject went through the same gait pattern, i.e. the movement was set to a frequency of 0.25 Hz for internal/external rotation.

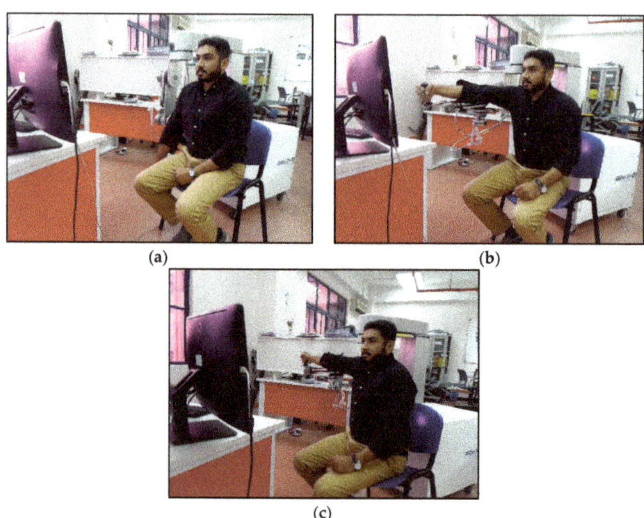

(a)

(b)

(c)

Figure 15. A subject performing exercise with assistance of exoskeleton device.

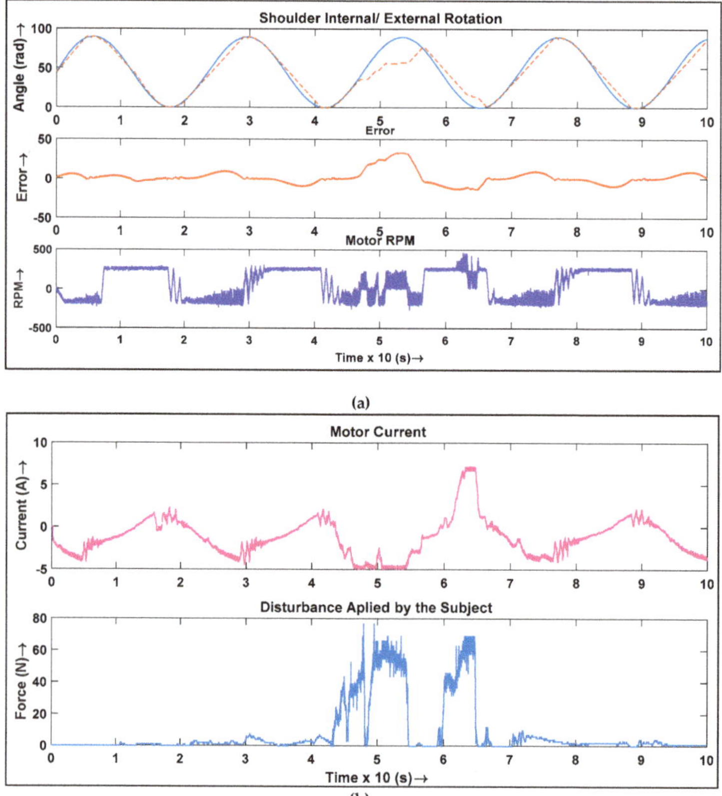

Figure 16. (**a**) Shoulder Internal/External Rotation, (**b**) Current driven from the motor and Disturbance applied by first subject.

Figure 16b shows the current required by the motor to move the shoulder to a desired angle. It also represents the amount of external force exerted by the patient on the system. The external force acts as an external disturbance to the system and its effect can be seen in the Figure 16b. As the external force to resist the movement increases, the motor current drastically increases to overcome the effect of disturbance. The maximum current drawn by the motor is 5.27 A when the maximum external resisting force of 68 N is applied to the joint.

The second exercise, namely shoulder abduction/adduction, was performed by the second subject; the bounds are set from 0° to 90°. Figure 17a represents the response of the system during shoulder abduction and adduction. It also represents the difference between the reference signal and response, which demonstrates a minute overshoot in the response. The gait pattern for this exercise is similar to that for the previous exercise, where the movement is set to a frequency of 0.25 Hz.

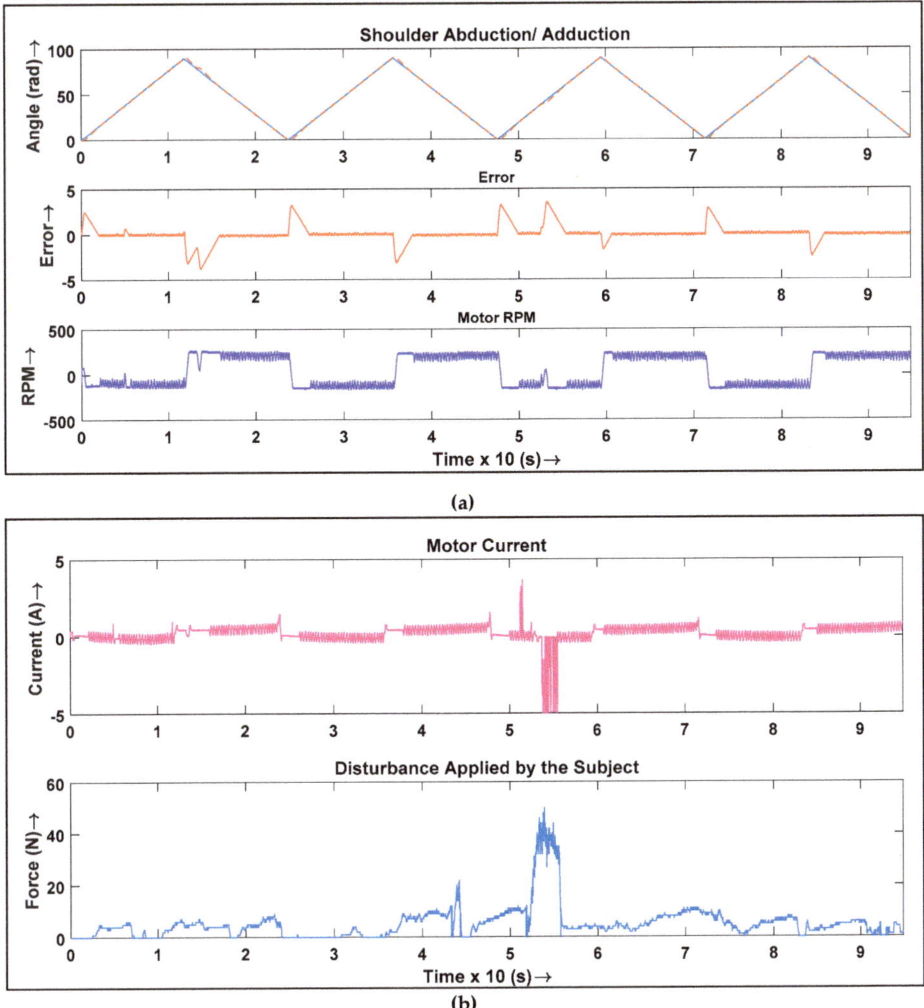

Figure 17. (a) Shoulder Abduction/ Adduction, (b) Current driven from the motor and Disturbance applied by second subject.

Figure 17b represents current drawn by the motor to achieve the target angle and the external force applied by the subject. When external resistance was exerted on the system by the subject, an unusual activity in the current was detected to overcome the external disturbance. A maximum current of 5.14 A was observed with 50N of applied force. The impact of this resistance can also be observed in Figure 17a, where disruption occurs when external force is applied while achieving the reference.

The third exercise, namely, shoulder extension/flexion, was performed by the third subject; the range of motion was fixed to oscillate between 0° to 90°. In this experiment a pulsating input was fed as a reference (see Figure 18).

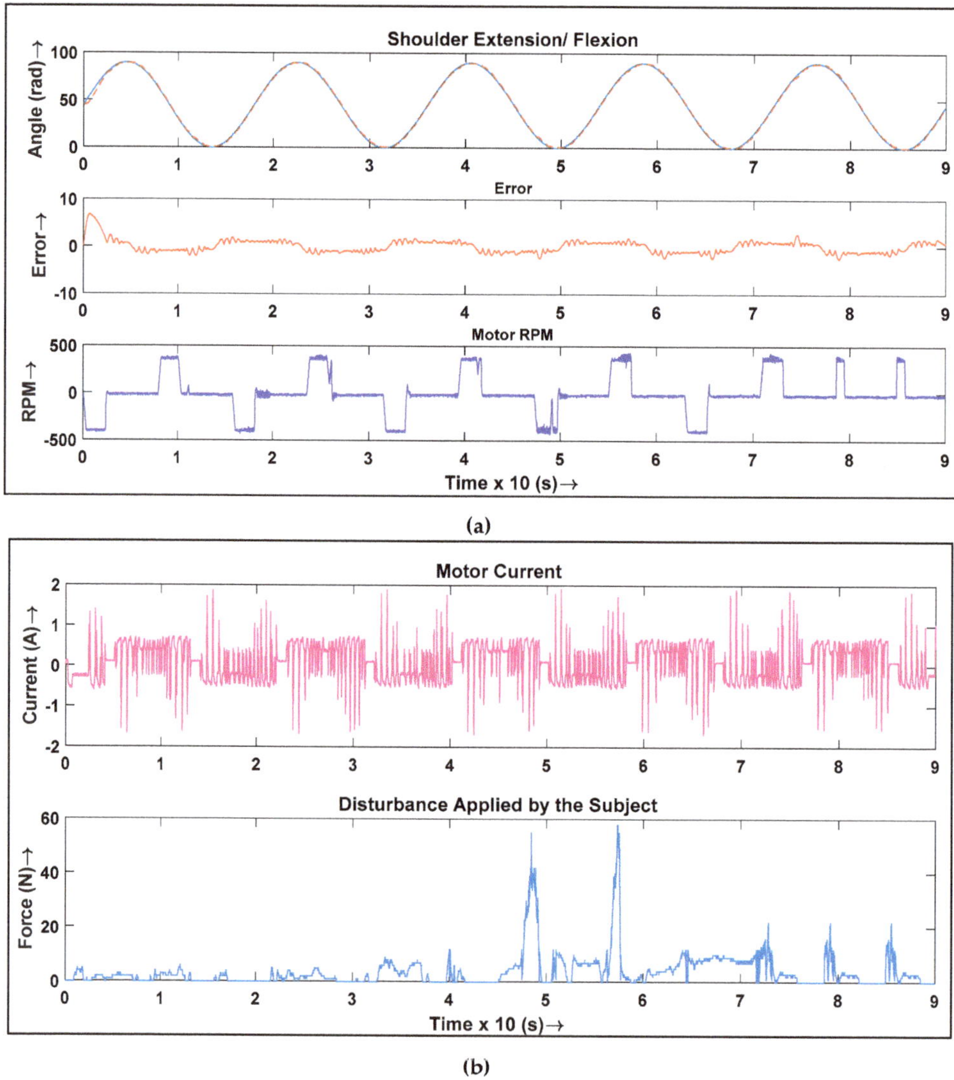

(a)

(b)

Figure 18. (**a**) Shoulder Extension/Flexion, (**b**) Current driven from the motor and Disturbance applied by third subject.

Figure 18a shows the motion tracking of the shoulder joint while performing extension/flexion. It also shows the error between reference and actual angle with the speed of the motor. In this exercise, each subject went through same gait pattern i.e. the exoskeleton is set to move with the frequency of 0.25 Hz for extension/flexion. Figure 18b shows the current driven by the motor to perform the exercise and force exerted by the patient to resist the movement. It can be seen that the motor draws more current than usual where the external force is applied; its impact on the movement can also be observed in Figure 18a. It is evident from the graph that a small amount of overshoot with no steady state was observed. The maximum force applied by the subject was recorded at 55N, while 4.5A of current was drawn by the motor.

5.4. Comparison of the PID Controllers Optimized with ABC and the Zeigler-Nichols Method

An experiment was performed to compare PID tuned controllers tuned with ABC and the Zeigler-Nichols method under the effect of disturbance. The parameters set for both techniques are similar to those defined in the simulation setup. The experiment was performed with the extension/flexion movement and the range of motion for both techniques was the same. The response of the system using ABC optimization and the Zeigler-Nichols method is shown in Figure 19.

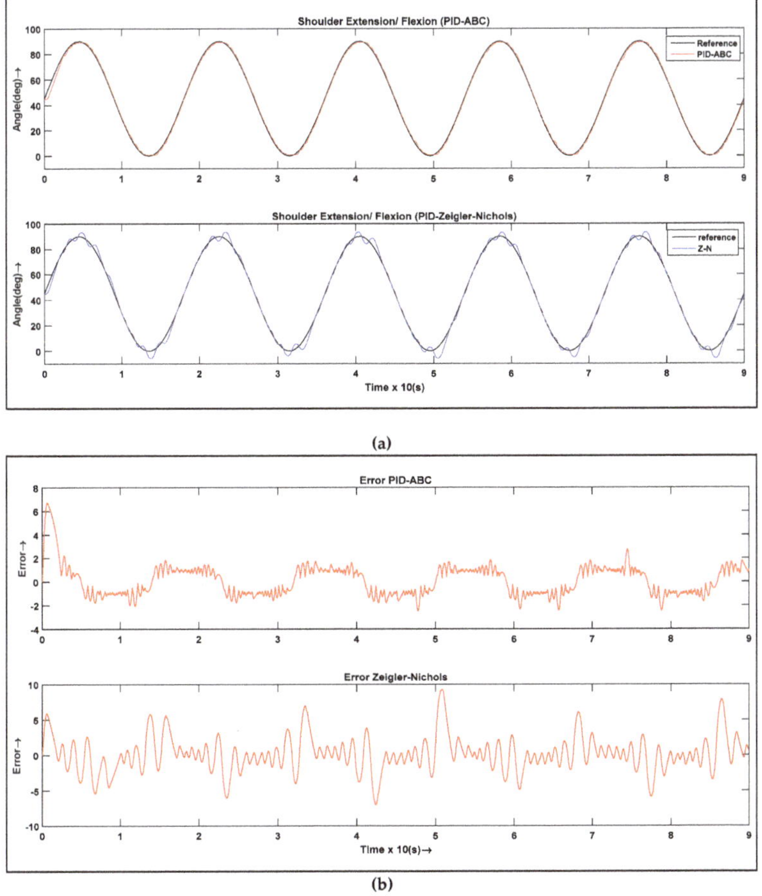

Figure 19. (a) Response of the system with ABC optimized PID and tuned using Zeigler-Nichols, (b) Error of the system with ABC optimized PID and tuned using Zeigler-Nichols.

It can be seen from Figure 19 that PID controller tuned with the Ziegler-Nichols has a low settling time and quick response. However, the response of the system is not smooth and has a steady state error. At the same time, ABC-optimized PID response is slower due to a greater settling time. However, it has a smooth response with virtually no steady-state error. The angular rotation of the arm is set to move from 0° to 90°. Figure 19a illustrates the response of the system tuned with ABC and the Zeigler-Nichols method, while the error signal is shown in Figure 19b. It is evident from the graph that the rise time of ZN-PID is lower than ABC-PID. The overshoot and steady state produced by ZN-PID are also larger than those of ABC-PID. Zeigler-Nichols based PID produces an angular drift of 3.64° while it is 0.63° for ABC-PID. A slower response is favorable for the rehabilitation system used in this study. The presence of large overshoot and steady state in the system is dangerous, as it may cause dislocation or fracture of the arm. Hence, it can be concluded that PID controller works better when optimized with ABC compared to PSO or Zeigler-Nichols method as ABC-PID produces very low overshoot, slower rise time and no steady state.

6. Conclusions

Rehabilitation robotics have been studied for decades, but few researchers have ever considered using optimization techniques for gait training. This paper presents PSO- and ABC-based tuning techniques for a 2-DOF PID controller used in the robot controlling trajectories of different exercise movements performed by the shoulder joints, namely internal/external rotation, abduction/adduction, and extension/flexion. RAX-1 was used as a mechanical experimental platform. The control parameters of the PID controller were tuned using the ABC and PSO algorithms as well as the Zeigler-Nichols method. The simulation results demonstrated better feasibility of the proposed controller in terms of robustness. An ABC-optimized PID controller was also implemented into the hardware for three subjects, with each performing different exercises. It was necessary for the robot to perform steady motion with no steady state error during the process of rehabilitation, as any abrupt movement could dislocate the shoulder joint. The hardware results showed that the controller could trace the desired trajectory with a very minute overshoot in the response and significantly low response times, which are desirable in rehabilitation. Finally, this study compared the performance of PID controllers optimized with ABC and the Zeigler-Nichols method, respectively. The controller tuned using the Zeigler-Nichols method demonstrated a decent rise time but large overshoot in the response, which is dangerous for rehabilitation applications. However, the hardware response of the system had less overshoot and no steady-state error when the ABC optimizer was used to tune the PID parameters. In summary, this study focused on finding optimal parameters of the PID controller used in an upper limb rehabilitation robotic system. In the future, we plan to broaden the spectrum of this study by incorporating various protocols related to the elbow and wrist rehabilitation with an updated mechanical structure. Analysis with other notable optimization techniques such as firefly and ant colony algorithms will be further investigated to compare their parameters and validate their capabilities in rehabilitation applications.

Author Contributions: K.N. and S.F.A. conceived the presented idea to use optimization of controller parameters for rehabilitation robotic system. M.K.J. and Y.R. implemented the idea and drafted this manuscript. M.M.B. and A.A. massively contributed in developing the mathematical model for robot exoskeleton. The project has been supervised by K.K. and Z.M.Y.

Funding: This research has been generously supported by a research grant by the Ministry of Science, Technology and Innovation (MOSTI) under the Program Flagship DSTIN (FP0514D003-2) for the development of a new technology identified as "Building Our Robotic Competitiveness in Medical Healthcare: Development of Robots for Assisted Recovery and Rehabilitation".

Acknowledgments: We would like to pay our gratitude to Universiti Kuala Lumpur for providing great research environment. The authors are also grateful to Technology Park Malaysia for their assistance in developing the mechanical structure for robot exoskeleton.

Conflicts of Interest: The authors declare no conflict of interest.

References

1. Page, S.J.; Levine, P.; Leonard, A.C. Modified constraint-induced therapy in acute stroke: A randomized controlled pilot study. *Neurorehabilit. Neural Repair* **2005**, *19*, 27–32. [CrossRef] [PubMed]
2. Williams, G.R. Incidence and characteristics of total stroke in the United States. *BMC Neurol.* **2001**, *1*, 2. [CrossRef]
3. Members, W.G.; Go, A.S.; Mozaffarian, D.; Roger, V.L.; Benjamin, E.J.; Berry, J.D.; Blaha, M.J.; Dai, S.; Ford, E.S.; Fox, C.S. Heart disease and stroke statistics—2014 update: A report from the American Heart Association. *Circulation* **2014**, *129*, e28.
4. Bingül, Z.; Karahan, O. A Fuzzy Logic Controller tuned with PSO for 2 DOF robot trajectory control. *Expert Syst. Appl.* **2011**, *38*, 1017–1031. [CrossRef]
5. Bharat, S.; Ganguly, A.; Chatterjee, R.; Basak, B.; Sheet, D.K.; Ganguly, A. A Review on Tuning Methods for PID Controller. *Asian J. Converg. Technol.* **2019**, *3*, 17–22.
6. Ali, A.; Ahmed, S.F.; Kadir, K.A.; Joyo, M.K.; Yarooq, R.S. Fuzzy PID controller for upper limb rehabilitation robotic system. In Proceedings of the 2018 IEEE International Conference on Innovative Research and Development (ICIRD), Bangkok, Thailand, 11–12 May 2018; pp. 1–5.
7. Kudinov, Y.; Kolesnikov, V.; Pashchenko, F.; Pashchenko, A.; Papic, L. Optimization of Fuzzy PID Controller's Parameters. *Procedia Comput. Sci.* **2017**, *103*, 618–622. [CrossRef]
8. Leng, G.; McGinnity, T.M.; Prasad, G. Design for self-organizing fuzzy neural networks based on genetic algorithms. *IEEE Trans. Fuzzy Syst.* **2006**, *14*, 755–766. [CrossRef]
9. Juang, C.-F.; Lu, C.-M.; Lo, C.; Wang, C.-Y. Ant colony optimization algorithm for fuzzy controller design and its FPGA implementation. *IEEE Trans. Ind. Electron.* **2008**, *55*, 1453–1462. [CrossRef]
10. Das, M.T.; Dulger, L.C.; Das, G.S. Robotic applications with particle swarm optimization (pso). In Proceedings of the 2013 International Conference on Control, Decision and Information Technologies (CoDIT), Hammamet, Tunisia, 6–8 May 2013; pp. 160–165.
11. Karaboga, D. *An Idea Based on Honey Bee Swarm for Numerical Optimization*; Technical Report-tr06; Erciyes University, Engineering Faculty, Computer Engineering Department: Kayseri, Turkey, 2005.
12. Iruthayarajan, M.W.; Baskar, S. *Optimization of PID Parameters Using Genetic Algorithm and Particle Swarm Optimization*; Curran Associates: New York, NY, USA, 2007.
13. Elbeltagi, E.; Hegazy, T.; Grierson, D. Comparison among five evolutionary-based optimization algorithms. *Adv. Eng. Inform.* **2005**, *19*, 43–53. [CrossRef]
14. Shen, Z.; Zhou, J.; Gao, J.; Song, R. Fuzzy Logic Based PID Control of a 3 DOF Lower Limb Rehabilitation Robot. In Proceedings of the 2018 IEEE 8th Annual International Conference on CYBER Technology in Automation, Control, and Intelligent Systems (CYBER), Tianjin, China, 9–13 July 2018; pp. 818–821.
15. Wu, J.; Huang, J.; Wang, Y.; Xing, K.; Xu, Q. Fuzzy PID control of a wearable rehabilitation robotic hand driven by pneumatic muscles. In Proceedings of the 2009 International Symposium on Micro-NanoMechatronics and Human Science, Nagoya, Japan, 8–11 November 2009; pp. 408–413.
16. Wu, J.; Wang, Y.; Huang, J.; Xing, K. Artificial Bee Colony algorithm based Auto-Disturbance Rejection Control for rehabilitation robotic arm driven by PM-TS actuator. In Proceedings of the 2012 International Conference on Modelling, Identification & Control (ICMIC), Hubei, China, 24–26 June 2012; pp. 802–807.
17. Tarmizi, W.F.W.; Elamvazuthi, I.; Perumal, N.; Nurhanim, K.; Khan, M.A.; Parasuraman, S.; Nandedkar, A. A Particle Swarm Optimization-PID controller of a DC Servomotor for Multi-Fingered Robot Hand. In Proceedings of the 2016 2nd IEEE International Symposium on Robotics and Manufacturing Automation (ROMA), Ipoh, Malaysia, 25–27 September 2016; pp. 1–6.
18. Annisa, J.; Darus, I.M.; Tokhi, M.; Mohamaddan, S. Implementation of PID Based Controller Tuned by Evolutionary Algorithm for Double Link Flexible Robotic Manipulator. In Proceedings of the 2018 International Conference on Computational Approach in Smart Systems Design and Applications (ICASSDA), Kuching, Malaysia, 15–17 August 2018; pp. 1–5.
19. Yan, F.; Wang, Y.; Xu, W.; Chen, B. Time delay control of cable-driven manipulators with artificial bee colony algorithm. *Trans. Can. Soc. Mech. Eng.* **2018**, *42*, 177–186. [CrossRef]
20. Bingul, Z.; Karahan, O. Comparison of PID and FOPID controllers tuned by PSO and ABC algorithms for unstable and integrating systems with time delay. *Optim. Control Appl. Methods* **2018**, *39*, 1431–1450. [CrossRef]

21. Begum, K.G.; Rao, A.S.; Radhakrishnan, T. Maximum sensitivity based analytical tuning rules for PID controllers for unstable dead time processes. *Chem. Eng. Res. Des.* **2016**, *109*, 593–606. [CrossRef]

22. Biodex, Manual, System 3 and 4 Upper Extremity Hemiparetic Attachments—Operation. Available online: https://www.biodex.com/physical-medicine/products/dynamometers/system-4-accessories/ue-hemiparetic-attachments (accessed on 3 December 2018).

23. Shah, J.A.; Rattan, S.; Nakra, B. End-effector position analysis using forward kinematics for 5 DOF pravak robot arm. *IAES Int. J. Robot. Autom.* **2013**, *2*, 112. [CrossRef]

24. Lai, L.-C.; Chang, Y.-C.; Jeng, J.-T.; Huang, G.-M.; Li, W.-N.; Zhang, Y.-S. A PSO method for optimal design of PID controller in motion planning of a mobile robot. In Proceedings of the 2013 International Conference on Fuzzy Theory and Its Applications (iFUZZY), Taipei, Taiwan, 6–8 December 2013; pp. 134–139.

25. Yousof, Z.M.; Billah, M.M.; Joyo, K.M.; Kadir, K.; Qisti, M. *Milestone 5 Report: Robotic System Development, Integration and Testing Product Testing*; Under the Project of Building Our Robotic Competitiveness in Medical Healthcare: Development of Robot for Assisted Recovery and Rehabilitation; Technology Park Malaysia Engineering Sdn Bhd: Kuala Lumpur, Malaysia, 2018.

26. Czarkowski, D.; O'Mahony, T. Intelligent controller design based on gain and phase margin specifications. In Proceedings of the Irish Signals and Systems Conference 2004, Belfast, Ireland, 30 June–2 July 2004.

27. Aouf, A.; Boussaid, L.; Sakly, A. A PSO algorithm applied to a PID controller for motion mobile robot in a complex dynamic environment. In Proceedings of the 2017 International Conference on Engineering & MIS (ICEMIS), Monastir, Tunisia, 8–10 May 2017; pp. 1–7.

28. Karaboga, D.; Basturk, B. On the performance of artificial bee colony (ABC) algorithm. *Appl. Soft Comput.* **2008**, *8*, 687–697. [CrossRef]

29. Kumar, A.; Kumar, D.; Jarial, S. A Review on Artificial Bee Colony Algorithms and Their Applications to Data Clustering. *Cybern. Inf. Technol.* **2017**, *17*, 3–28. [CrossRef]

30. Mustafa, A.M.; Al-Saif, A. Modeling, simulation and control of 2-R robot. *Glob. J. Res. Eng.* **2014**, *14*, 49–54.

31. Naidu, K.; Mokhlis, H.; Bakar, A.A. Multiobjective optimization using weighted sum artificial bee colony algorithm for load frequency control. *Int. J. Electr. Power Energy Syst.* **2014**, *55*, 657–667. [CrossRef]

 *electronics*

Article

RTPO: A Domain Knowledge Base for Robot Task Planning

Xiaolei Sun, Yu Zhang * and Jing Chen

College of Intelligence Science and Technology, National University of Defense Technology, Changsha 410073, China; sunxiaoleijc@163.com (X.S.); chenjingnudt@163.com (J.C.)
* Correspondence: redarmy_zy@163.com; Tel.: +86-139-7585-6486

Received: 21 August 2019; Accepted: 26 September 2019; Published: 1 October 2019

Abstract: Knowledge can enhance the intelligence of robots' high-level decision-making. However, there is no specific domain knowledge base for robot task planning in this field. Aiming to represent the knowledge in robot task planning, the Robot Task Planning Ontology (RTPO) is first designed and implemented in this work, so that robots can understand and know how to carry out task planning to reach the goal state. In this paper, the RTPO is divided into three parts: task ontology, environment ontology, and robot ontology, followed by a detailed description of these three types of knowledge, respectively. The OWL (Web Ontology Language) is adopted to represent the knowledge in robot task planning. Then, the paper proposes a method to evaluate the scalability and responsiveness of RTPO. Finally, the corresponding task planning algorithm is designed based on RTPO, and then the paper conducts experiments on the basis of the real robot TurtleBot3 to verify the usability of RTPO. The experimental results demonstrate that RTPO has good performance in scalability and responsiveness, and the robot can achieve given high-level tasks based on RTPO.

Keywords: ontology; robot task planning; knowledge base; knowledge representation

1. Introduction

Nowadays, artificial intelligence (AI) has become a growing field in which many theory studies and practical applications are enjoying a boom. Deep learning, a typical approach to AI, has propelled its entry into a new stage of development [1]. In the field of robotics, AI technology represented by deep learning has also grown to be extensive and vital in several applications [2,3]. However, the deep learning model, a kind of end-to-end learning, helps to achieve uninterpretable and opaque results, which limits its application in some areas requiring knowledge reasoning. For example, in the field of robot combat task planning, the plans need to be interpretable, so that the operational commander can evaluate the advantages and disadvantages of the plans.

Since the 1970s, AI researchers have gradually realized that symbolic knowledge representation is exerting a key role in more powerful AI systems, believing that knowledge and reasoning is the core of AI. From then on, ontology has been developing robustly as a form of knowledge base. It can represent and understand the sophisticated world. Until now, it has been widely employed in various fields such as AI [4,5], Semantic Web [6,7], Information Science [8], etc. Ontology-based task planning is essentially a series of relevant queries and reasoning on ontology knowledge [9]. Made up of a large number of individuals, concepts, and their semantic relations, it can first interpret the elements of the query and reasoning path of the task planning process on ontology knowledge [10]. Similar to a human's thinking mode, a robot can also utilize knowledge and knowledge reasoning to realize smart high-level decision-making.

The challenge in building RTPO is how to efficiently and reasonably represent the intricate task knowledge. There is a need for thorough consideration of temporal and spatial information, as well as continuous and discrete information. Furthermore, it needs to have good performance in scalability and responsiveness, so as to guarantee the usability in the task planning algorithm.

The world knowledge, which is categorized into static knowledge, can be easily and feasibly exhibited, as is said by the previous researches on robot task planning. In this regard, as a result of the feasibility of representing world knowledge, the problem file of the task planner can be obtained from the world knowledge efficiently [11,12]. However, comparatively, the causal knowledge [13], is more represented in the formal language, like the PDDL (Planning Domain Definition Language) [14], HTN (Hierarchical Task Network) [15], and so on. Correspondingly, as is shown in Figure 1, the domain file of the task planner is inclined to come from the causal knowledge manually, which makes it comparatively complex and lacking universality in large-scale applications. Consequently, the paper proposed a task planning algorithm based on the built RTPO, avoiding the complex process of manual generation of domain file.

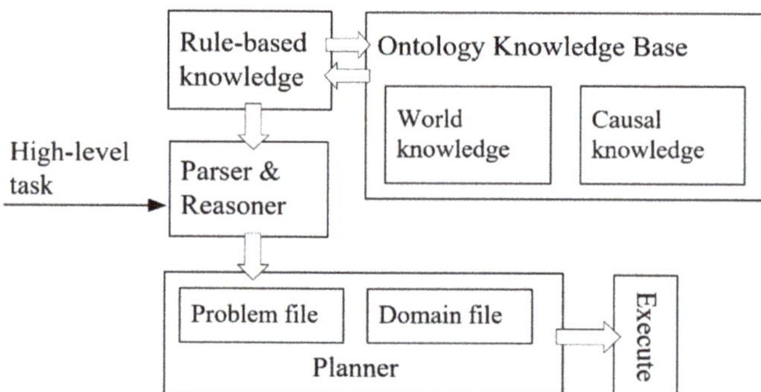

Figure 1. Knowledge representation in robot task planning.

The representations of high-level tasks and atomic actions in RTPO are independent from each other. Then, task planning is realized by matching the execution preconditions of atomic actions and their effects on the environment from the initial state to the goal state. In this way, with the changing of input tasks, the task planning module can also be carried out in accordance with the present atomic action resources. The plans generated by the task planning algorithm will add and update to RTPO, and will improve the efficiency of task planning if the same task needs to be planned next time.

The research work mainly involves two aspects. Firstly, an ontology knowledge base is built for robot task planning and a method to evaluate the knowledge base is proposed. Secondly, an experiment with an indoor study case is carried out to verify the usability of RTPO and the flexibility of the proposed task planning algorithm. During the real robot experiments, we used the ROS (Robot Operating System) [16] under Ubuntu as the software system, and adopted TurtleBot3 as the hardware platform. The experimental results demonstrate that RTPO is of good usability and the task planning algorithm harbors flexibility to address the unexpected events.

There are some typical study works and applications on robot knowledge base. KnowRob [17–19], an integrated knowledge management system for autonomous robots, aims at the construction of an indoor service robot knowledge base. It is composed of ontology, entities, and extensible reasoning engine in OWL language. However, KnowRob holds the limitation of having less abundant knowledge. RoboEarth [20–22], based on KnowRob, has already given definition to sub-actions of specific tasks, along with the definition of the temporal and spatial constraints among different sub-actions. That is, when a high-level task needs to be planned, the user should request the high-level task to generate its corresponding plans. As the task planning methods have to rely on specific high-level task instructions, it is possible that task planning will fail to input other different task instructions.

Besides this, ORO (OpenRobots Common-Sense ontology) [23,24] is built in OWL and stored in the OpenJena ontology management library. The knowledge information is reasoned by the Pellet reasoner. Nonetheless, the ORO knowledge management system highlights the interaction between robots and humans. SWARMs (Smart and Networking Underwater Robots in Cooperation Meshes) Ontology [25] is built to represent and understand the knowledge of unmanned underwater robots to facilitate cooperation between them. The SWARMs ontology in Reference [13] is divided into four domain-specific ontologies (environmental model, vehicle model, communication model, and mission model) and a core ontology which connects the four domain-specific ontologies. The Semantic Web Rule Language (SWRL) is adopted in the SWARMs ontology for the compensation of the inability of OWL to represent complex rule formations and relations. Reference [26] offers an evaluation of the SWARMs Ontology and verification of applicability by conducting multiple underwater robots cooperation experiments. Comparatively, the ontology in SWARMs is merely constructed for the unmanned underwater robots, so it is hard to extend to other robot application types.

The ontology theory is also applied to other fields, such as industrial production collaboration [27] and navigation in indoor scenarios [28,29]. The work in References [27,30] portrays the domain knowledge for the robot task planning in logic language BC [31,32]. However, this approach is disadvantageous in sharing knowledge and updating the new knowledge inferred. In addition, the ontology representation language OWL [33,34] is endowed with sharing as its natural advantage, which can be easily shared on the web.

Currently, the existing robot ontology knowledge bases embrace the complete representation, which covers each expectation in robot task planning. However, the actual application of the ontology is lacking in intelligence. Meanwhile, the knowledge is just stored in libraries and queried to get the output in use. That is, different knowledge and reasoning lacks associations to obtain new knowledge, like humans' thinking mode.

The main contributions of this paper can be summarized as follows:

- It contributes a domain knowledge base RTPO for robots to have a better understanding of the task planning knowledge.
- An evaluation method of knowledge base is proposed and implemented to test scalability and responsiveness in this paper.
- A task planning algorithm based on RTPO is proposed which has good flexibility and avoids the shortcoming of manual editing domain knowledge in traditional task planners.
- We carried out the experimental research and applied the proposed approach on the real robot.

The rest of the paper is arranged as follows. Section 2 introduces the purpose and requirements for the building of RTPO. Section 3 presents the building method of RTPO. Section 4 describes the knowledge representation in RTPO. In Section 5, we implement a method to test the scalability and responsiveness of RTPO and an algorithm of robot task planning based on RTPO on a real robot. Finally, Section 6 summarizes all the work and introduces future work.

2. Building Considerations for Robot Task Planning Ontology (RTPO)

The building considerations for RTPO will be presented in this section. The first part will illustrate the purpose of RTPO building, and then the requirements for RTPO building will be unfolded.

2.1. Purpose of Robot Task Planning Ontology (RTPO)

It is essential to clarify the purpose of RTPO building. Based on a clear understanding of the purpose, it is possible to decide the contents to be included in the RTPO, which expectations they should be divided into, and which tools should be adopted. The RTPO is mainly associated with knowledge in connection to the robot task planning, including the robot-itself-related concepts, the concepts of environment, and the task-related concepts. Our RTPO design and building aim to provide a comprehensive and available knowledge base for the application of robot task planning. Various heterogeneous robots can query and reason the knowledge from the knowledge base to obtain useful information, which helps to improve the efficiency of task planning and increase the intelligence by reasoning knowledge to plan automatically.

2.2. Requirements of Robot Task Planning Ontology (RTPO)

There are many approaches and forms for building the ontology, but there is no unified standard pattern. However, the great ontology should harbor the standard features which should not change with the different approaches and forms. Therefore, the RTPO should meet a set of requirements to ensure an appropriate outcome. Generally, the paper takes into consideration that the great ontology in robotics should cover the following characteristics [35]:

- With unambiguous knowledge representation, it is easy to be understood by humans and robots.
- With strong editability, it is easy to be operated and utilized by developers.
- With knowledge representation, it is consistent and free from contradictory knowledge or definitions.

3. The Building of Robot Task Planning Ontology (RTPO)

This section will give a detailed description of the building method of RTPO, composed of the model, approaches, and knowledge reasoning.

3.1. The Model of Robot Task Planning Ontology (RTPO)

This section starts with a formal definition of the model in RTPO employed for building the RTPO later on.

Definition 1. *A high-level task model is defined as a 5-tuple.*

$$\mathcal{T} = \left(T_{name}, T_{attr}, T_{entity}, T_{tasker}, T_{methods}\right) \tag{1}$$

Consisting of a set of task names T_{name}, a set of task attributes T_{attr}, e.g., the start time and initial state, a set of entities carrying out the tasks T_{entity}, a set of taskers giving the high-level tasks T_{tasker}, and a set of methods to decompose the high-level task $T_{methods}$.

Definition 2. *an atomic action model is defined as a 5-tuple.*

$$\mathcal{A} = \left(A_{name}, A_{attr}, A_{entity}, A_{pre}, A_{effect}\right) \tag{2}$$

Consisting of a set of atomic actions names A_{name}, a set of atomic actions attributes A_{attr}, a set of entities carrying out the atomic actions A_{entity}, a set of execution preconditions of atomic actions A_{pre}, and a set of actions effects A_{effect}.

3.2. The Approaches to Build Robot Task Planning Ontology (RTPO)

Some software editors are available to edit the ontologies, such as Protégé [36], Neon-tool kit, OntoWiki, and so on [37]. The editor employed for RTPO building is Protégé, equipped with the RDF (Resource Description Framework) triple, which was developed by Stanford University. Besides this, Java is applied as the development language of Protégé. With many embedded plugins in Protégé, the set has been evolved to be one of the essential ontology editors [36]. The Protégé harbors many conception constraints, which helps to add and update the corresponding inferred knowledge. In addition, the knowledge of robot task planning can be derived from Internet, books, or manual editing by humans, just as is shown in Figure 2.

In Figure 2, the query of relevant knowledge from RTPO is realized in the language SWI–Prolog [38]. This logic language, which can be well combined with ROS, can easily and quickly query the relevant knowledge from RTPO. Also, the rules and knowledge can be edited and stored through SWI–Prolog language. The matching query of knowledge in task planning algorithm later can be conducted through rules.

In the duration of specific implementation, the knowledge is stored in .owl and .pl file formats. Specifically, the .owl file stores the ontology knowledge in RDF triples and the .pl file stores the rules knowledge in Prolog language. RTPO harbors the application in which the paper adopts the ROS as the robot middleware, which is featured by the ROS node and communication through the mechanism of the topic.

OWL, whose file can be generated by Protégé, is further applied as logic language in our research work. An OWL file's generation comes following the ontology building, which holds excellent portability, and can be well used in other approaches. It also breaks down the knowledge interaction barrier among different knowledge systems. OWL file can be published in the World Wide Web and may refer to or be referred from other OWL files [34].

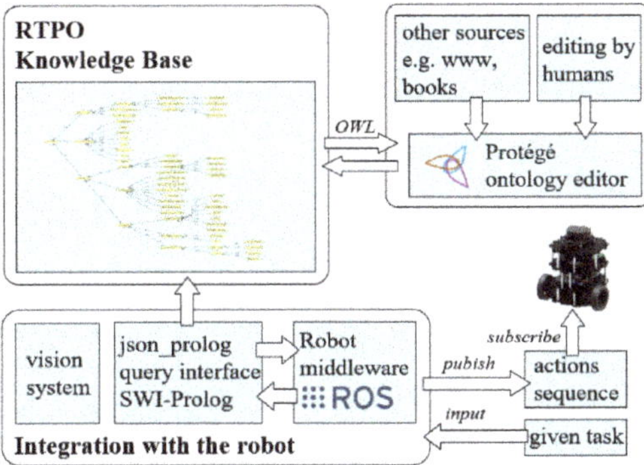

Figure 2. The implementation process of RTPO building.

ROS provides the rosprolog function package, through which the users can conveniently explore and debug the knowledge in the terminal window. However, if you would like to apply the knowledge in your robot's control program, you need a way to send queries from your program. This functionality is provided by the json_prolog package (http://wiki.ros.org/json_prolog). It provides a service that exposes a Prolog shell via the ROS node. The ROS Node programs can be written in multiple languages, such as Python, C++, and Java. Moreover, the ROS provides a number of function packages for developers. A snapshot of the hierarchy of RTPO is shown in Figure 3.

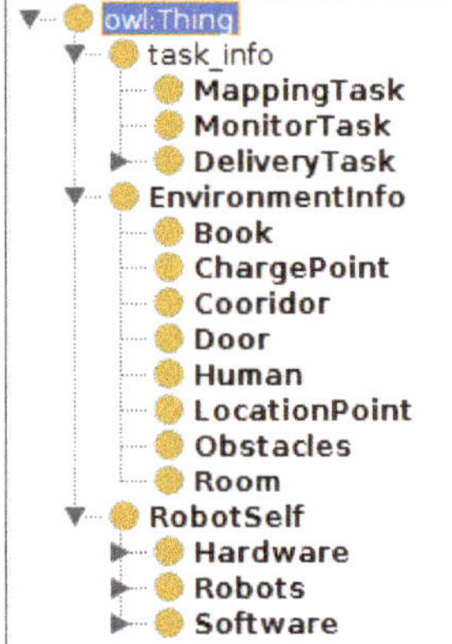

Figure 3. A snapshot of the hierarchy of the Robot Task Planning Ontology (RTPO).

4. Knowledge Representation in Robot Task Planning Ontology (RTPO)

This section offers the introduction of the knowledge representation in RTPO we have built, including the knowledge types, knowledge structure, and knowledge reasoning in RTPO.

4.1. The Knowledge in Robot Task Planning Ontology (RTPO)

During the decision process of robots, the task planning module receives the high-level task and subsequently generates a sequence of atomic actions. The execution of atomic actions sequence will have an effect on the environment state, changing the environment state until the goal state is achieved. It is evident that the process of task planning requires knowledge. The knowledge for robot task planning can be composed of two parts. One is the knowledge related to the initial state of the world, which also corresponds to the problem file of task planners. The other is the knowledge relevant to how the given high-level task can be planned from the initial state to the goal state which is also corresponding to the domain file of task planners. The first kind of knowledge exists in the form of world knowledge, including information of environment and robots themselves. The second kind of knowledge is presented in causal knowledge, including the preconditions and the effects on environment state of atomic actions. Taking the high-level task "DeliveryHandbooktoLeo" as an example, the robot task planner calls for the location and capability of robots, the location of humans, the environmental map, and so on, which is defined as world knowledge. Besides this, what the robot needs to know is how to search the optimal sequence of atomic actions from the initial state to the goal state. For example, the task decomposition includes the following steps of "DeliveryHandbooktoLeo": move to Jack, get the handbook, move to Leo, and give the book to Leo. That is the causal knowledge [13].

Similar to a human, when a robot wants to complete a task, first of all, it has to know who it is. That refers to the robot-itself knowledge, including the hardware and software, location information, dynamics information, and so forth. Secondly, in order to complete the task, it also has to harbor the familiarity with the surrounding environment, namely environmental knowledge, which consists of the location and recognition of humans and objects, environment map, information of other robots, and so on. Finally, following the assignment, it needs to understand how to decompose the given high-level task into atomic actions and how to re-plan when the environment is changed, which is the task knowledge.

Therefore, it is easy to understand that nothing else is more important than the three kinds of knowledge representation in our research work. Thus, the paper classifies the RTPO building into three parts: robot ontology, environment ontology, and task ontology, respectively. Next, the paper will describe the structure of RTPO in detail.

4.2. The Structure of Robot Task Planning Ontology (RTPO)

RTPO contains three parts: robot ontology, environment ontology, and task ontology. The task ontology describes the knowledge connected with robot tasks, such as the task decomposition, task allocation, and task execution, etc. Then, the robot ontology is designed to portray the knowledge or concepts corresponding to the robot itself in the hierarchical structure. Finally, environment ontology gives a description of the knowledge relevant to the environment, such as the environmental map, environmental objects, and so on. The overall structure diagram of the RTPO is shown in Figure 4.

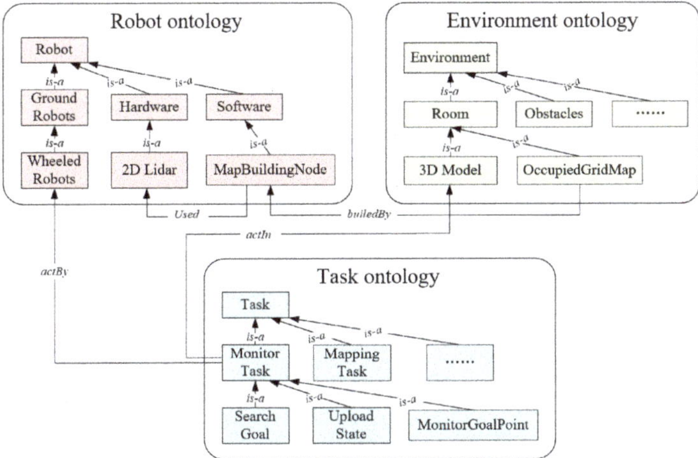

Figure 4. The diagram of the whole structure of Robot Task Planning Ontology (RTPO).

4.2.1. Robot Ontology

As shown in Figure 5, the robot ontology contains the following three parts: robots, hardware, and software, which demonstrate the capability and characteristics of robots. Robots include various types of robot, like ground robots, underwater robots, and air robots. The concepts of robots can be installed as individuals in Protégé. The hardware is composed of the components and devices the various types of robots may have. It can be further divided into perception devices, navigation devices, and base devices. Navigation devices refer to the hardware devices that robots need to navigate and locate, such as IMU (Inertial Measurement Unit). Perception devices refer to the hardware devices that robots need to perceive and understand the environment, such as lidar and camera. Base devices refer to the hardware devices related to the low-level control of the robot, such as motor, battery, and so on. Software consists of the functional ROS node, which can publish its specific topic and subscribe to the other topics. It can also achieve communication among control nodes. A variety of relationships can be defined by developers to describe the relationship among the robot ontology and other ontologies.

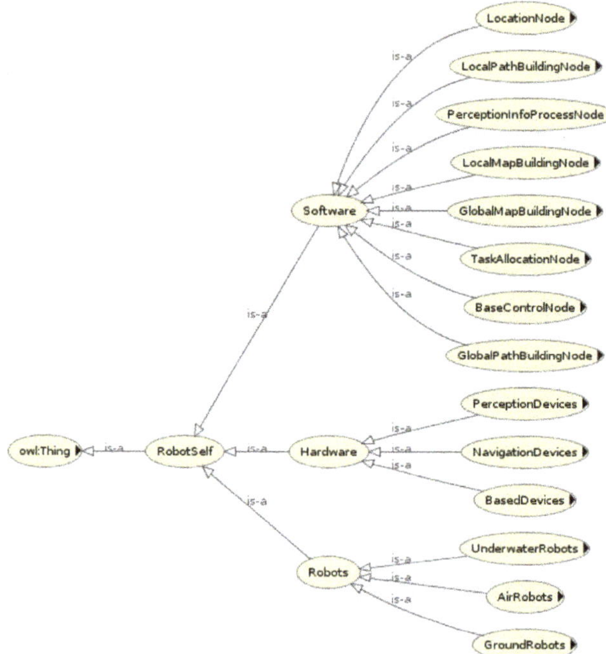

Figure 5. The robot ontology shown in Protégé.

4.2.2. Environment Ontology

Environment ontology is mainly designed for the detailed description of the indoor environment where TurtleBot3 moves. According to the experiment environment, it is seen in Figure 6 that the environment ontology includes the map, obstacles, doors, and other objects in an indoor environment. Besides this, the research requires that the environment ontology should be instantiated partly, for instance, the door and the room. The RTPO is capable of adding and updating environment knowledge after the task planning of robots.

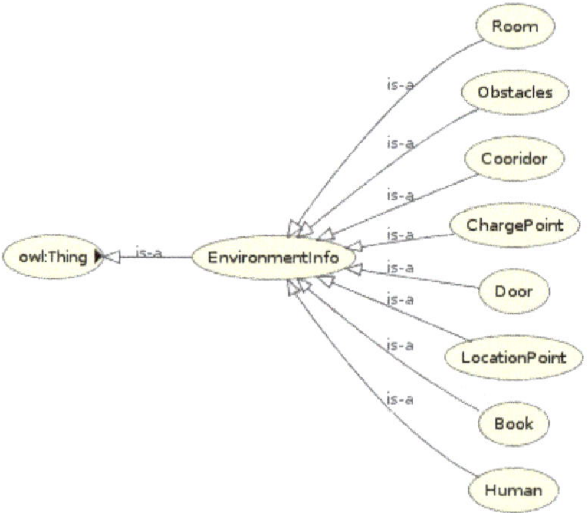

Figure 6. The environment ontology shown in Protégé.

4.2.3. Task Ontology

As is displayed in Figure 7, the paper extends the task ontology from four typical tasks, which are, respectively, monitor task, mapping task, delivery task, and charge task. The domain files which most planners use for decomposition are written manually, which is inefficient and less portable. The task description structure is similar to the HTN. Thus, we can decompose the given high-level tasks through ontology for task planning. For better utilizing the hierarchical structures, the representation method of task ontology is designated to meet the experimental requirements. Going forward, the preconditions of each task and sub-action are defined, including the robot's capability and environment state. Besides this, the atomic actions are defined to have effects (delete and add) on the environment state.

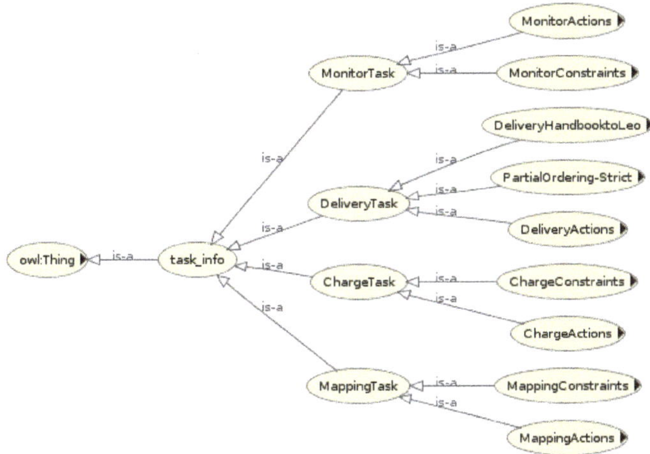

Figure 7. The task ontology shown in Protégé.

Figure 8 interprets the representation method of atomic actions, defined as the smallest granular actions which can be directly executed by a robot. Atomic actions are made up of execution preconditions and action effects. The former will have an effect on environment state, such as deleting or adding some state, as shown in Figure 8.

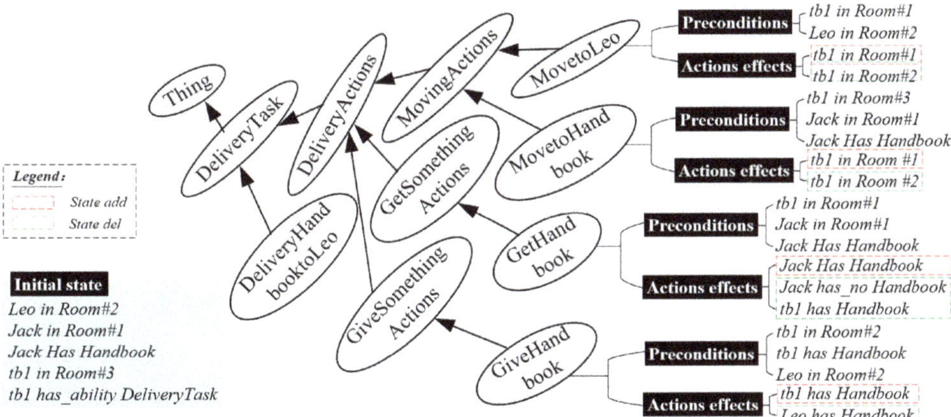

Figure 8. The representation method of atomic actions.

Generally speaking, for a given high-level task, for example, DeliveryHandbooktoLeo in Figure 9, its representation information in the ontology is insufficient in the initial state, and lacks the update of reasoning knowledge, such as the constraints among different atomic actions. The task knowledge is represented with the insufficient form as follows at the initial state.

Class: DeliveryHandbooktoLeo
    SubClassOf:
        DeliveryTask
(subAction some GetHandbook)
and (subAction some GiveHandbookToLeo)
and (subAction some MovetoHandbook)
and (subAction some MovetoLeo)

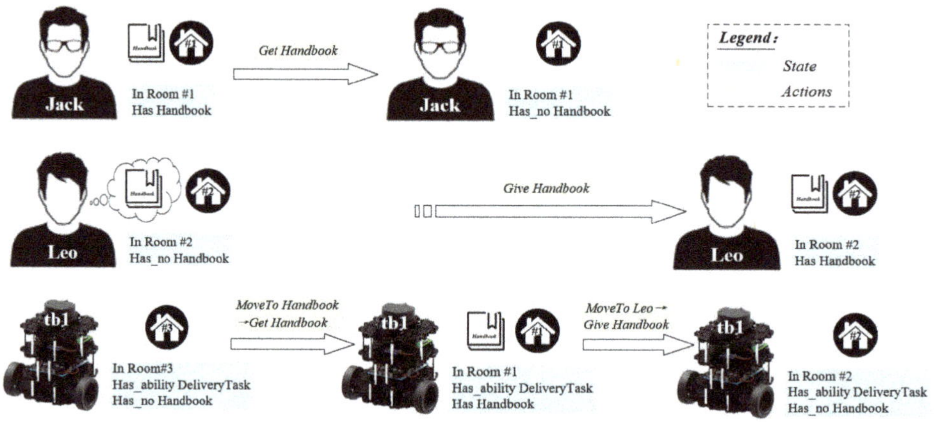

Figure 9. The environment state changes in a simple indoor service DeliveryHandbooktoLeo: the TurtleBot3 need to get the handbook from Jack and then give the handbook to Leo.

The robot can adopt a specific task planning algorithm on the basis of the initial state of environment. The task planning algorithm aims to match the preconditions and action effects of atomic actions, so that the execution order constraints can be generated and added into the task ontology. The system can further obtain the atomic action sequence of the specific task. At this moment, new knowledge is obtained and updated based on the reasoning of existing knowledge.

After the task planning algorithm runs, a complete representation of a high-level task decomposition can be obtained and shown as follows, composed of the parent classes, the subactions, and the execution order constraints among subactions. When the same specific task requires planning, the users can query the ontology to obtain the atomic actions sequence of the specific task directly, which will improve the efficiency of task planning.

Class: DeliveryHandbooktoLeo
    SubClassOf:
        DeliveryTask
(subAction some GetHandbook)
and (subAction some GiveHandbookToLeo)
and (subAction some MovetoHandbook)
and (subAction some MovetoLeo)
and (orderingConstraints value DeliveryActions12)
and (orderingConstraints value DeliveryActions13)
and (orderingConstraints value DeliveryActions14)

and (orderingConstraints value DeliveryActions23)
and (orderingConstraints value DeliveryActions24)
and (orderingConstraints value DeliveryActions34)

Relying on the individual class in Protégé, the order constraints among sub-actions is defined as follows. Assuming that a specific task has n sub-actions, it is easy to know that the total number to define the specific task is C_n^2, provided that we completely define the execution order of all sub-actions.

Individuals: DeliveryActions12
    Types:
        PartialOrdering-Strict
    Annotations:
        occursAfterInOrdering GetHandbook
        occursBeforeInOrderingMovetoHandbook

4.2.4. Communications Among the Three Parts

Apart from the contents above, corresponding communications also exist among the three parts which connect this knowledge with another. These relationships can be defined according to the developers' own needs. Taking the indoor service monitor task as an example (Figure 10), the monitor task is achieved by the mobile wheeled robot TurtleBot3 and in the environmental map of Room2. The latter is built by the Lidar of TurtleBot1. Accordingly, these three ontology modules can be linked and added to constraints so that they make up the whole ontology jointly.

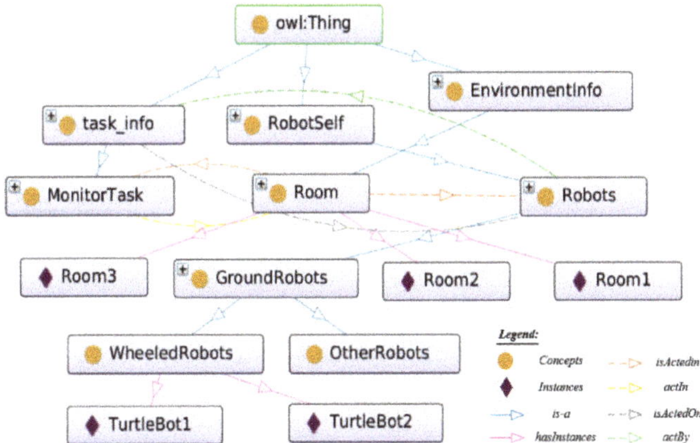

Figure 10. The communications among three parts in Protégé.

4.3. Knowledge Reasoning

Figure 11 gives an example of the reasoning in the RTPO knowledge base. The left is the clarification of knowledge and the relationship between the knowledge, but the position of the book is not indicated. That is, we fail to achieve the position of the handbook just through the existing ontology knowledge. On the right, after adding the following rule knowledge, we can infer that the exact location of the handbook is room #1.

In (Book, Room) :-
    Has (Human, Book),
    In (Human, Room)

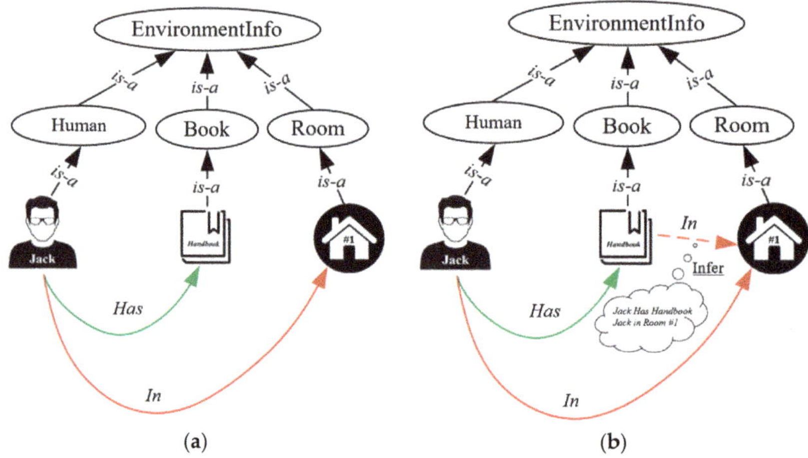

Figure 11. Example of knowledge reasoning. (**a**) Left: the handbook is asserted to be just a book, and Jack has it. (**b**) Right: If some rule knowledge is added, the knowledge system can infer the location of the handbook.

5. Evaluation and Experiments

The evaluation of robot knowledge base is a complicated work. So far, there is neither an established benchmark nor evaluation methods. In addition, each robot knowledge system has different coverage and application fields, which brings great challenges to the evaluation of robot knowledge base. It is difficult to evaluate robot knowledge base with only one or a few indicators. Therefore, considering the knowledge system and its application field, this section evaluates RTPO with a combinatorial evaluation method. We adopt a quantitative evaluation method for the scalability and responsiveness of RTPO. Then, this section conducts an experimental case study to verify the usability of RTPO and the flexibility of task planning algorithm based on RTPO on TurtleBot3.

5.1. The Evaluation of Robot Task Planning Ontology (RTPO)

The scalability of robot knowledge system lies in the efficiency of knowledge updating and storage. The addition of new knowledge is the premise of the application expansion of the knowledge base. The good scalability of knowledge base is the basis of its sustainable development. In order to test the scalability of RTPO, we designed a test algorithm in the process of knowledge base instantiation. The process of knowledge base instantiation refers to the update of individual knowledge, such as the addition of a cabinet and a chair in the indoor environment. Therefore, we designed an algorithm for writing new knowledge to RTPO, which can automatically generate a large number of individuals in a single loop statement:

g_individuals (0).
g_individuals (?Num) :-
    new_individuals (?Num).
    ?Num **is** ?Num−1.
    g_individuals (?Num).

The **g_individuals** (?Num) function just calls the **new_individuals** (?Num) functions Num times, which then generates Num individuals in RTPO. We used the prolog's **time**() that is ?-**time(g_individuals** (?Num)) to evaluate the performance of **g_individuals** (?Num). The scalability of RTPO is tested by changing the number of automatically generated individuals and then counting the time consumed by each number of individuals. Figure 12 shows how the consumed time changes with the number of generated individuals ((blue square markers). It is easy to see that the consumed time increases linearly with the number of generated individuals, where it takes about 2.34 s to generate 55,000 individuals. The maximum generation rate is about 23,500 individuals per second. As a comparison, KnowRob [18] has a maximum generation rate of 22,000 individuals per second and ORO [24] has a maximum generation rate of 7245 individuals per second.

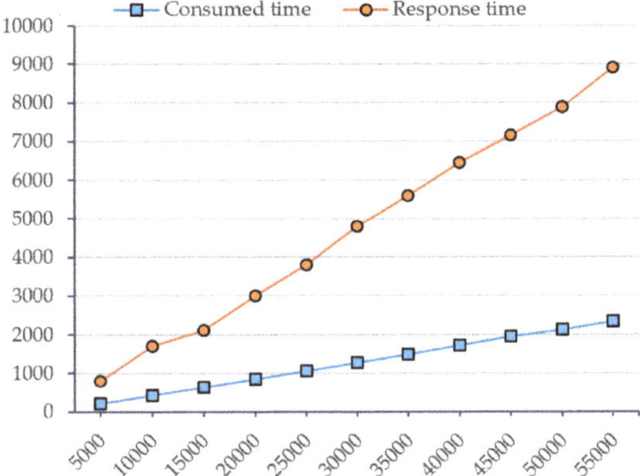

Figure 12. The consumed time changes with the number of generated instances (blue square markers) and the response time changes with the number of individual knowledge to query (yellow circle markers).

The responsiveness of knowledge system is mainly reflected in the knowledge query speed. The faster the knowledge query speed, the faster the responsiveness of knowledge system and the better the performance of knowledge system. We choose the Prolog statement "?- **time** (**findall**(?A, **owl_individual_of** (?A, 'Obstacles')))" to calculate the query speed of individuals knowledge. As shown in Figure 12 (yellow circle markers), the experimental results show that the response time increases linearly with the increase in the number of individual knowledge to query. Query 52,000 individuals can be completed within 10 s, which ensures the real-time performance of RTPO in the application. Compared with the KnowRob [18], both knowledge systems have similar responsiveness. In a word, RTPO has good scalability and responsiveness.

5.2. Verification Using a Case Study

In this section, we take the indoor delivery task as a case study to study the robot task planning based on RTPO on TurtleBot3.

5.2.1. Hardware and Software

As is shown in Figure 13, the hardware system in our system is established on TurtleBot3, which is a new generation of mobile robot platform established on ROS (Robot Operation System). The series of TurtleBot from TurtleBot1 to TurtleBot3 is becoming more and more powerful with the ROS. Based on ROS, it is an ideal and essential platform to do research work. Table 1 explains the configuration table of the burger.

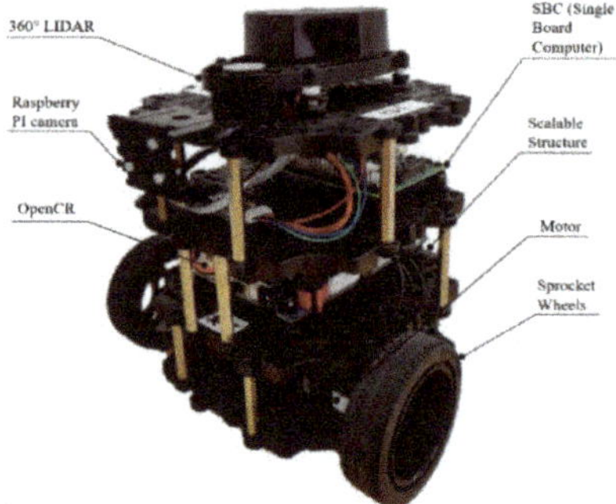

Figure 13. The TurtleBot3 Burger.

Table 1. The configuration table of burger.

Items	Configuration
LIDAR	360-degree laser LIDAR LDS-01 (HLS-LFCD2)
SBC	Raspberry PI 3 and Intel Joule 570x
Battery	Lithium polymer 11.1 V 1800 mAh
	Gyroscope 3 Axis
IMU	Accelerometer 3 Axis
	Magnetometer 3 Axis
MCU	OpenCR (32-bit ARM Cortex®M7)
Motor	DYNAMIXEL(XL430)

Accordingly, the software system is based on ROS, which is the most popular and vital middleware for robot system development. Figure 14 contributes our software system framework, which illustrates that the design and development of ROS nodes contributes to the central part of the software system framework. Based on ROS, the software system framework can be respectively classified into three control levels: control of actions, control of navigation, and control of velocity, corresponding to the task planning module, navigation module, and base control module.

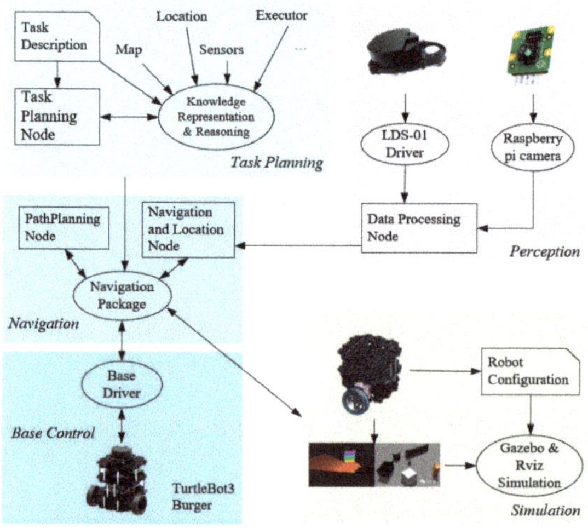

Figure 14. The framework of the software system.

5.2.2. The Experimental Scenario

An assumption is made in the case study that the robot is able to grasp objects. In addition, it is noteworthy that the adopted TurtleBot3 in the experiment only harbors mobility and obstacles perception. However, our research attaches importance to the decision-making at the related top level to the robot task planning and does not focus on such basic-level issues as how to control the movement and navigation. Therefore, we make a reasonable assumption that the robot is able to grasp objects.

Under the circumstances of the real robot experiment, the robot needs to get the handbook and then give it to Leo. The built experimental scenario is shown in Figure 15a. The object elements in the environment are displayed in virtue of the label objects on the ground, for example, persons, books and bookshelves. The environment map built through the LIDAR LDS–01 is shown in Figure 15b.

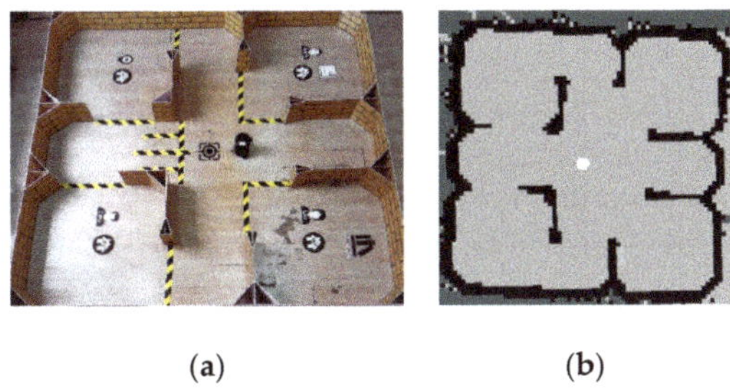

(**a**) (**b**)

Figure 15. The experimental environment we built: (**a**) the real experimental scenario; (**b**) the corresponding environment map built by TurtleBot3 in Gmapping algorithm.

During the verification experiment, the communication among the devices is made through LAN, which shares a master computer and topic to realize the communication requirements between devices. As shown in Figure 16, ontology knowledge is stored in computer PC_2. Besides this, computer PC_1 undertakes the master computer. It is also the central computer employed to run the master node. Turtlebot tb_1 commits the program storage of map building, navigation, and path planning. It is worth noting that the map building algorithm adopts Gmapping; the local path planning algorithm adopts the DWA (Dynamic Window Approach), the global path planning algorithm adopts the D*, and the location algorithm adopts the amcl.

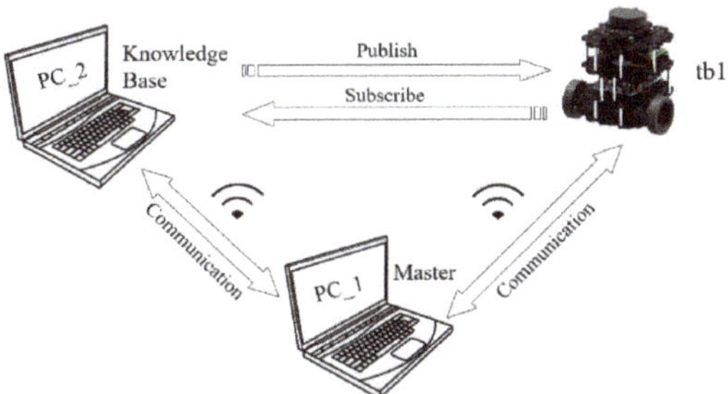

Figure 16. The communication among different devices.

5.2.3. The Experiments and Results

The proposed task planning algorithm Algorithm 1 based on RTPO is shown as the pseudo-code below. Its inputs are the initial state s, the given high-level task t, and the ontology knowledge O. The output of the algorithm is the sequence of atomic actions, which is also the plan for accomplishing the t from the initial state.

Algorithm 1 Task Planning Algorithm Based on Robot Task Planning Ontology (RTPO)

Input: s: the initial state; *t*: the given high-level task; *O*: the ontology knowledge
Output: *P*: A plan for accomplishing the *t* from the initial state;
1:   **procedure** generate a plan for accomplishing the *t*
2:        P = the empty plan
3:        function task_planning (t)
4:            if *t* is a primitive task then
5:                modify s by deleting del(t) and adding add(t)
6:                append t to P
7:            else
8:                for all subtask in subtasks(t) do
9:                    if preconditions(subtask) matches the s then
10:                        task_planning (subtask)
11:        return P
12:   **end procedure**

The specific implementation process of the study case on TurtleBot3 is demonstrated in Figure 17. The atomic actions sequence can be obtained by the task planning algorithm based on RTPO, as shown in Figure 18. The given high-level task "DeliveryHandbooktoLeo", whose initial environment state is "Leo in Room#2; Jack in Room#1; Jack Has Handbook; tb1 in Room#3; tb1 has_ability DeliveryTask", is decomposed into a sequence of atomic actions: MovetoHandbook, GetHandbook, MovetoLeo, and GiveHandbookToLeo. The corresponding action attributes, such as the target point and the time constraint of some actions, can be obtained by virtue of the analysis of the atomic action list and querying the RTPO. The robot then subscribes to the message and performs the corresponding actions through the navigation and path planning algorithm. Finally, the given high-level task is completed with achieving the goal environment state: "tb1 in Room#2; Leo in Room#2; Leo has Handbook".

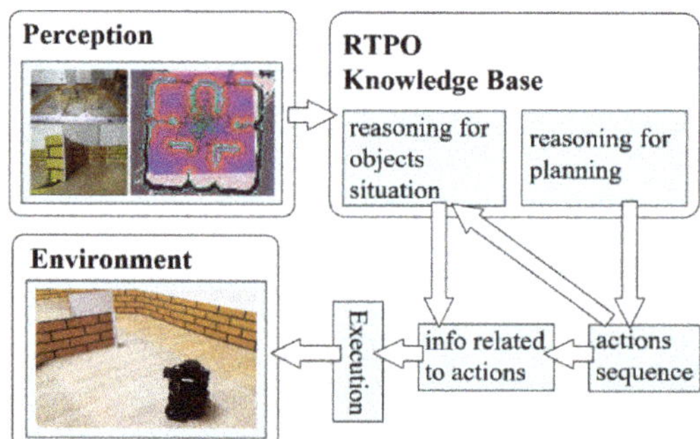

Figure 17. The specific implementation process of study case on TurtleBot3.

```
?- owl_parse('package://rtpo/owl/DeliveryTask.owl').
true.

?- plan_subactions(DeliveryTask:'DeliveryHandbooktoLeo', sub).
sub = [DeliveryTask:'MovetoHandbook',DeliveryTask:'GetHandbook',
DeliveryTask:'MovetoLeo',DeliveryTask:'GiveHandbookToLeo'].
```

Figure 18. The snapshot of the query sentence from the ROS terminal.

Figure 19 shows a sequence of snapshots for the real execution of the atomic action sequence generated by the decomposition of the given high-level task "DeliveryHandbooktoLeo" in a real-world experimental scenario. A sequence of atomic actions obtained from RTPO is displayed as navigating across the target points in a proper order according to atomic action's type. For example, the Moveto action needs to move to the specific target point. Figure 19a shows that the Turtlebot tb_1 is at the initial position. Figure 19b demonstrates that the Turtlebot tb_1 has subscribed the topic of atomic action MovetoHandbook and then the atomic action MovetoHandbook is executed. Figure 19c shows that the Turtlebot tb_1 executes the atomic action MovetoLeo. Figure 19d presents that the Turtlebot tb_1 has arrived at the position of Leo. The experimental video can be watched at the supplementary material Video S1. The experimental result demonstrates that the robot knowledge base RTPO is of good usability and can be well used in the robot task planning with the proposed task planning algorithm.

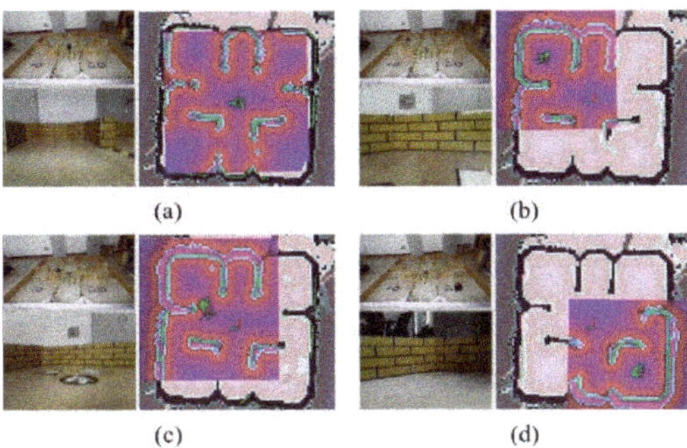

Figure 19. The sequence of snapshots for the execution of the task: "PutHandbookonBookshelf". (**a**) The initial position of TurtleBot3; (**b**) The atomic action of MovetoHandbook. (**c**) The atomic action MovetoLeo; (**d**) The arrival at the position of Leo.

Besides this, another test scenario is further designed and implemented to further validate the flexibility of task planning algorithm based on RTPO. In this scenario, we suppose that Leo will ask the TurtleBot3 to put the handbook on the bookshelf in room #2 after the robot completed the delivery task "DeliveryHandbooktoLeo". However, in accordance with the topic related to the battery published by the TurtleBot3, the battery of TurtleBot3 becomes insufficient at this time. In this regard, the robot makes a flexible plan for this scenario. First, it needs to recharge its battery in Room #3 and then put the handbook on the bookshelf Room #2, as shown in Figure 20. The given high-level task "PutHandbookonBookshelf" from Leo is decomposed into a sequence of atomic actions: GetHandbook, BatteryCharge, MovetoBookshelf, PutonHandbook. Figure 20e shows that the Turtlebot tb_1 moves to charge point to have a charge and Figure 20f demonstrates that the Turtlebot tb_1 moves to the target position of Bookshelf carrying the Handbook. The experimental video can be watched at supplementary material Video S2. The experimental results show that the task planning algorithm based on RTPO has good flexibility to address the unexpected events, such as the insufficient battery.

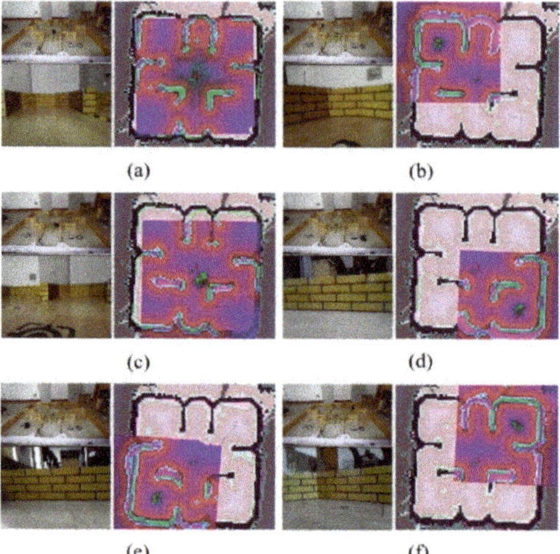

Figure 20. The sequence of snapshots for the execution of the task: "PutHandbookonBookshelf". (**a**) The initial position of TurtleBot3; (**b**) The atomic action of MovetoHandbook. (**c**) The atomic action MovetoLeo; (**d**) The arrival at the position of Leo; (**e**) The atomic action of BatteryCharge; (**f**) The action of MovetoBookshelf.

6. Conclusions

In conclusion, the paper manages to build a robot ontology called RTPO which is applied to the robots' task planning. Then, RTPO is designed and implemented followed by a proposed evaluation method to test its scalability and responsiveness. The test results show that RTPO has good performance of scalability and responsiveness compared with the existing knowledge base. Finally, we proposed a task planning algorithm based on RTPO and conducted the real robot experiments to verify the usability of RTPO and the flexibility of the proposed algorithm. The experimental results show that the robot can complete the given high-level task smoothly, and can also address unexpected events with good flexibility.

Future research work mainly focuses on the following aspects. Firstly, a large number of available robot ontologies have been constructed, such as KnowRob, ORO, SWARMs, and so on. In this aspect, it is suggested that more consideration of future research should be taken into the fusion of different ontologies. Then, multiple heterogeneous robots can cooperatively help to accomplish some more complex tasks which cannot be accomplished by just a single robot. Thereby, it is of great significance to study cooperative task planning and application based on multi-robot task planning. Additionally, the real world is complex and changeable, due to inaccuracy, randomness, and incompleteness. Therefore, it is necessary to study robot task planning under uncertain environments. Finally, the application of cloud-based knowledge will reduce robots' dependence of specific hardware. It is conducive to the research of multi-agent and swarm intelligence.

Supplementary Materials: The following are available online at https://susy.mdpi.com/user/manuscripts/displayFile/8ec60ffb629e5ad518f79653d69d980d/supplementary, Video S1: DeliveryHandbooktoLeo, Video S2: PutHandbookonBookshelf.

Author Contributions: Conceptualization, X.S. and Y.Z.; Formal analysis, X.S.; Funding acquisition, J.C.; Investigation, X.S.; Methodology, X.S. and Y.Z.; Project administration, Y.Z. and J.C.; Resources, J.C.; Supervision, J.C.; Validation, X.S.; Writing – original draft, X.S.; Writing – review & editing, Y.Z. and J.C.

Funding: This research was funded by the National Natural Science Foundation of China (Grant No. 61806212, No.61603403, No. U1734208 and No. 61702528).

Conflicts of Interest: The authors declare no conflict of interest.

References

1. Goodfellow, I.; Bengio, Y.; Courville, A. *Deep Learning*; MIT Press: Cambridge, MA, USA, 2016.
2. Yang, P.-C.; Suzuki, K.; Kase, K.; Sasaki, K.; Sugano, S.; Ogata, T. Repeatable folding task by humanoid robot worker using deep learning. *IEEE Robot. Autom. Lett.* **2016**, *2*, 397–403. [CrossRef]
3. Levine, S.; Pastor, P.; Krizhevsky, A.; Ibarz, J.; Quillen, D. Learning hand-eye coordination for robotic grasping with deep learning and large-scale data collection. *Int. J. Robot. Res.* **2018**, *37*, 421–436. [CrossRef]
4. Shiang, C.W.; Tee, F.S.; Halin, A.A.; Yap, N.K.; Hong, P.C. Ontology reuse for multiagent system development through pattern classification. *Softw. Pract. Exp.* **2018**, *48*, 1923–1939. [CrossRef]
5. El-Sappagh, S.; Alonso, J.M.; Ali, F.; Ali, A.; Jang, J.-H.; Kwak, K.-S. An ontology-based interpretable fuzzy decision support system for diabetes diagnosis. *IEEE Access* **2018**, *6*, 37371–37394. [CrossRef]
6. Liu, J.; Li, Y.; Tian, X.; Sangaiah, A.K.; Wang, J. Towards Semantic Sensor Data: An Ontology Approach. *Sensors* **2019**, *19*, 1193. [CrossRef]
7. Wen, Y.; Zhang, Y.; Huang, L.; Zhou, C.; Xiao, C.; Zhang, F.; Peng, X.; Zhan, W.; Sui, Z. Semantic Modelling of Ship Behavior in Harbor Based on Ontology and Dynamic Bayesian Network. *ISPRS Int. J. Geo-Inf.* **2019**, *8*, 107. [CrossRef]
8. Ibrahim, M.E.; Yang, Y.; Ndzi, D.L.; Yang, G.; Al-Maliki, M. Ontology-based personalized course recommendation framework. *IEEE Access* **2019**, *7*, 5180–5199. [CrossRef]
9. Jeon, H.; Yang, K.-M.; Park, S.; Choi, J.; Lim, Y. *An Ontology-Based Home Care Service Robot for Persons with Dementia*; IEEE: Piscataway, NJ, USA, 2018; pp. 540–545.
10. Xu, G.; Cao, Y.; Ren, Y.; Li, X.; Feng, Z. Network security situation awareness based on semantic ontology and user-defined rules for Internet of Things. *IEEE Access* **2017**, *5*, 21046–21056. [CrossRef]
11. Stock, S.; Mansouri, M.; Pecora, F.; Hertzberg, J. *Hierarchical Hybrid Planning in a Mobile Service Robot*; Springer: Berlin/Heidelberg, Germany, 2015; pp. 309–315.
12. Wang, Y.; Sun, H.; Chen, G.; Jia, Q.; Yu, B. Hierarchical task planning for multiarm robot with multiconstraint. *Math. Probl. Eng.* **2016**, *2016*, 2508304. [CrossRef]
13. Galindo, C.; Fernández-Madrigal, J.-A.; González, J.; Saffiotti, A. Robot task planning using semantic maps. *Robot. Auton. Syst.* **2008**, *56*, 955–966. [CrossRef]
14. Cashmore, M.; Fox, M.; Long, D.; Magazzeni, D.; Ridder, B.; Carrera, A.; Palomeras, N.; Hurtos, N.; Carreras, M. Rosplan: Planning in the robot operating system. In Proceedings of the Twenty-Fifth International Conference on Automated Planning and Scheduling, Jerusalem, Israel, 7–11 June 2015.
15. Lu, F.; Tian, G.; Li, Q. Autonomous cognition and planning of robot service based on ontology. *Jiqiren/Robot* **2017**, *39*, 423–430.
16. IsaacSaito.Wiki: ROS [EB/OL]. Available online: http://wiki.ros.org/ROS/ (accessed on 23 April 2013).
17. Tenorth, M. Knowledge Processing for Autonomous Robots. Ph.D. Thesis, Technische Universität München, Munich, Germany, 2011.
18. Tenorth, M.; Beetz, M. KNOWROB—Knowledge processing for autonomous personal robots. In Proceedings of the IEEE/RSJ International Conference on Intelligent Robots and Systems, St. Louis, MO, USA, 11–15 October 2009; pp. 4261–4266.
19. Tenorth, M.; Beetz, M. Representations for robot knowledge in the KnowRob framework. *Artif. Intell.* **2015**, *247*, 151–169. [CrossRef]
20. Tenorth, M.; Perzylo, A.C.; Lafrenz, R.; Beetz, M. Representation and Exchange of Knowledge about Actions, Objects, and Environments in the RoboEarth Framework. *IEEE Trans. Autom. Sci. Eng.* **2013**, *10*, 643–651. [CrossRef]
21. Waibel, M.; Beetz, M.; Civera, J.; d'Andrea, R.; Elfring, J.; Galvez-Lopez, D.; Häussermann, K.; Janssen, R.; Montiel, J.M.M.; Perzylo, A.; et al. Roboearth—A world wide web for robots. *IEEE Robot. Autom. Mag. (RAM)* **2011**, *18*, 69–82. [CrossRef]
22. Riazuelo, L.; Civera, J.; Montiel, J.; Montiel, J.M.M. C2tam: A cloud framework for cooperative tracking and mapping. *Robot. Auton. Syst.* **2014**, *62*, 401–413. [CrossRef]
23. Lemaignan, S. Grounding the Interaction: Knowledge Management for Interactive Robots. *KI-Künstliche Intell.* **2013**, *27*, 183–185. [CrossRef]

24. Lemaignan, S.; Ros, R.; Mösenlechner, L.; Alami, R.; Beetz, M. ORO, a knowledge management platform for cognitive architectures in robotics. In Proceedings of the 2010 IEEE/RSJ International Conference on Intelligent Robots and Systems, Taipei, Taiwan, 18–22 October 2010; pp. 3548–3553.

25. Li, X.; Bilbao, S.; Martín-Wanton, T.; Bastos, J.; Rodriguez, J. SWARMs ontology: A common information model for the cooperation of underwater robots. *Sensors* **2017**, *17*, 569. [CrossRef]

26. Landa-Torres, I.; Manjarres, D.; Bilbao, S.; Del Ser, J. Underwater robot task planning using multi-objective meta-heuristics. *Sensors* **2017**, *17*, 762. [CrossRef]

27. Sadik, A.R.; Urban, B. An Ontology-Based Approach to Enable Knowledge Representation and Reasoning in Worker–Cobot Agile Manufacturing. *Future Internet* **2017**, *9*, 90. [CrossRef]

28. Diab, M.; Akbari, A.; Din, M.U.; Rosell, J. PMK—A Knowledge Processing Framework for Autonomous Robotics Perception and Manipulation. *Sensors* **2019**, *19*, 1166. [CrossRef]

29. Schlenoff, C.; Prestes, E.; Madhavan, R.; Goncalves, P.; Li, H.; Balakirsky, S.; Kramer, T.; Miguelanez, E. An IEEE standard ontology for robotics and automation. In Proceedings of the 2012 IEEE/RSJ International Conference on Intelligent Robots and Systems, Algarve, Portugal, 7–12 October 2012; IEEE: Piscataway, NJ, USA, 2012; pp. 1337–1342.

30. Khandelwal, P.; Zhang, S.; Sinapov, J.; Leonetti, M.; Thomason, J.; Yang, F.; Gori, I.; Svetlik, M.; Khante, P.; Lifschitz, V.; et al. Bwibots: A platform for bridging the gap between ai and human–robot interaction research. *Int. J. Robot. Res.* **2017**, *36*, 635–659. [CrossRef]

31. Khandelwal, P.; Yang, F.; Leonetti, M.; Lifschitz, V.; Stone, P. Planning in Action Language BC while Learning Action Costs for Mobile Robots. In Proceedings of the Twenty-Fourth International Conference on Automated Planning and Scheduling, Portsmouth, NH, USA, 21–26 June 2014.

32. Lee, J.; Lifschitz, V.; Yang, F. Action Language BC: Preliminary Report. In Proceedings of the Twenty-Third International Joint Conference on Artificial Intelligence, Beijing, China, 3–9 August 2013; pp. 983–989.

33. McGuinness, D.L.; Van Harmelen, F. OWL web ontology language overview. *W3C Recomm.* **2004**, *10*, 2004.

34. OWL Working Group. OWL—Semantic Web Standard [EB/OL]. Available online: https://www.w3.org/2001/sw/wiki/OWL,2-013-12-21 (accessed on 21 August 2019).

35. Zhai, Z.; Ortega, J.-F.M.; Martínez, N.L.; Castillejo, P. A Rule-Based Reasoner for Underwater Robots Using OWL and SWRL. *Sensors* **2018**, *18*, 3481. [CrossRef] [PubMed]

36. TaniaTudorache. Protégé Wiki [EB/OL]. Available online: https://protegewiki.stanford.edu/wiki/Main_Page (accessed on 23 May 2016).

37. Maedche, A.; Staab, S. Ontology learning for the Semantic Web. *Intell. Syst. IEEE* **2001**, *16*, 72–79. [CrossRef]

38. Clocksin, W.F.; Mellish, C.S. *Programming in Prolog*; Springer: Berlin/Heidelberg, Germany, 1981.

 *electronics*

MDPI

Article

Automatic Spray Trajectory Optimization on Bézier Surface

Wei Chen [1,2]**, Junjie Liu** [1]**, Yang Tang** [3,*] **and Huilin Ge** [1]

[1] School of Electronics and Information, Jiangsu University of Science and Technology, Zhenjiang 212003, China; cwchenwei@aliyun.com (W.C.); junjiel_mtr@163.com (J.L.); gehuilin404@163.com (H.G.)
[2] School of Automation, Southeast University, Nanjing 210096, China
[3] School of Science, Jiangsu University, Zhenjiang 212013, China
* Correspondence: ty800117@ujs.edu.cn

Received: 26 November 2018; Accepted: 18 January 2019; Published: 1 February 2019

Abstract: The trajectory optimization of automatic spraying robot is still a challenging problem, which is very important in the whole spraying work. Spray trajectory optimization consists of two parts: spray space path and end-effector moving speed. A large number of spraying experiments have proved that it is very important to find the best initial trajectory of spraying. This paper presents an automatic spray trajectory optimization that is based on the Bézier surface. Spray the workpiece for Bezier triangular surface modeling and find the best initial trajectory of the spraying robot, establish the appropriate spraying model, plan the appropriate space path, and finally plan the trajectory optimization along the specified painting path. The validity and practicability of the method presented in this paper are proved by an example. This method can also be extended to other applications.

Keywords: initial trajectory; trajectory optimization; Bézier surface

---

1. Introduction

With the development of intelligence, multi-intelligence systems have received extensive attention and application. It can be seen that intelligent automation is widely used in painting robots [1,2]. Surface modeling of the sprayed workpiece is the first step in the trajectory optimization of the spray painting robot and the key to designing the spray path of the spray painting robot. At present, there are two main types of surface modeling methods of sprayed workpiece for off-line programming system of spray painting robot: (1) Surface modeling method based on CAD (Computer Aided Design) model. The modeling method based on the CAD model is that the CAD model data of the workpiece have been obtained before the surface modeling, and the spraying path of the spray painting robot can be planned according to the CAD model of the workpiece. (2) Surface modeling method based on the workpiece scanning system. If there is no CAD model data for a workpiece, or if the surface shape of actual workpiece does not match the CAD model data, then the workpiece needs to be scanned to obtain its new CAD data. The surface of the workpiece is approximated by a simple plane, sphere, cylinder, or other parametric surface, allowing spray path planning on these parametric surfaces.

In the previous work, we focused on the trajectory optimization of complex curved surface. In the reference [3], the Surface modeling method based on the CAD model is adopted. The CAD model data of the workpiece is obtained, and can plan the spraying path of the spraying robot according to the CAD model of the workpiece. The trajectory optimization of spray painting robot for complex curved surface based on the exponential mean Bézier method is proposed. The advantage is that it does not need to split the complex curved surface.

During the off-line programming operation for spray painting robot, after the surface modeling for the workpiece is finished, the following work is to optimize the trajectory of spray painting robot [4,5].

Since the point on the trajectory is a six-dimensional vector in the Cartesian coordinate system, it is very complicated to express it in mathematical expressions and the solving process is very difficult. Therefore, the general idea of trajectory optimization for spray painting robot is usually as follows: First, find the spatial path of the spray painting robot on the workpiece surface, and then find out the optimal time sequence along the specified spatial path. That is, the consistency of the paint thickness on the workpiece surface is the highest and the spray painting time is the shortest at what speed the end effector sprays along the specified spatial path. According to this idea, the optimized trajectory for spray painting consists of two parts: the spatial path of the spray painting operation and the moving speed of the end effector.

On the other hand, a large number of spray painting experiments have shown that the uniformity of the paint thickness can be significantly improved. That is, the spray painting effect can be improved by optimizing the initial trajectory of the spray painting robot in the initial stage of spray painting [6–8]. In other words, finding the best initial trajectory for spray painting is critical to the further trajectory optimization. Therefore, the trajectory optimization for spray painting robot can be divided into the following four steps: (1) Finding the optimal initial trajectory of spray painting robots. (2) According to the geometric features of sprayed workpiece surface, establish a suitable spray painting model. (3) Plan an appropriate spatial path for spray painting. (4) Plan trajectory optimization along the specified painting path.

The initial trajectory selection, the establishment of spraying model and path planning are all the bases of trajectory optimization. This research is based on the assumption that the workpiece CAD model was not acquired in advance. The innovation is that the Bézier triangular surface modeling method is adopted under the CAD model data without the workpiece. Firstly, the Bézier surface is analyzed and the method for searching the optimal initial trajectory of the spray painting robot is given according to the features of the Bézier triangular surface. Subsequently, the spray painting model of Bézier surface is established, and the mathematic expression of paint thickness at a certain point on the Bézier surface is given. Finally, the optimized trajectory of Bézier surface is obtained by using the ideal point method in the mathematical programming with the uniformity of paint thickness and the shortest spray painting time as optimization objectives along the specified spray painting path. The advantage of this method is that a good initial path of automatic spraying is determined at the beginning of the spraying process, which can significantly provide uniformity of coating thickness, that is, improve the spraying effect.

2. Bézier Triangular Surface Modeling Method of Sprayed Workpiece

Since the Bernstein polynomial has many superior properties, it is widely used in parametric polynomial curve surfaces of many forms. Based on the features of the sprayed workpiece surface, the Bézier triangular surface is constructed by using the Bernstein polynomial as the basis function.

Definition 1. *There is an arbitrary given triangle on the plane whose vertices T_1 T_2, T_3 are in the counterclockwise direction. Point P is any point in the plane where the triangle $T_1T_2T_3$ is located.*

Subsequently, we define:

$$u_1 = \frac{[PT_2T_3]}{[T_1T_2T_3]}, u_2 = \frac{[T_1PT_3]}{[T_1T_2T_3]}, u_3 = \frac{[T_1T_2P]}{[T_1T_2T_3]} \tag{1}$$

In the Equation (1), $[T_1T_2T_3]$ represents the directed area of the triangle $T_1T_2T_3$; When T_1, T_2, T_3 is counterclockwise, $[T_1T_2T_3]$ represents the area of the triangle $T_1T_2T_3$, that is $[T_1T_2T_3] = S$; When T_1, T_2, T_3 is clockwise, $[T_1T_2T_3]$ represents the opposite number of the area of the triangle, that is, $[T_1T_2T_3] = -S$. Afterwards, we call (u_1, u_2, u_3) as the area coordinate of Point P, recorded as $P = (u_1, u_2, u_3)$. Triangle $T_1T_2T_3$ is also called as coordinate triangle, as shown in Figure 1.

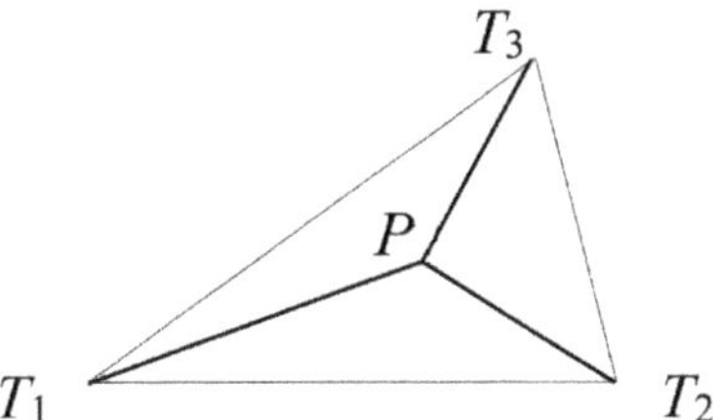

Figure 1. The Area Coordinate of Point P on the Coordinate Triangle.

Definition 2. *Suppose the area coordinate of point P on the coordinate triangle T is (u_1, u_2, u_3), and then we define:*

$$B^n_{i,j,k}(P) = \frac{n!}{i!j!k!} u_1^i u_2^j u_3^k, \quad i + j + k = n \tag{2}$$

As the Bernstein basis function $((n+1)(n+2)/2$ in all), They have the following properties:

(1) Non-negative: $B^n_{i,j,k}(P) \geq 0, P \in T, i + j + k = n$;

(2) Normative: $\sum\limits_{i+j+k=n} B^n_{i,j,k}(P) = 1$;

where, according to the triangular theorem we can have:

$$(a + b + c)^n = \sum\limits_{i+j+k=n} \frac{n!}{i!j!k!} a^i b^j c^k, \quad a, b, c \in R, \quad n \in N \tag{3}$$

For any P, $u_1 + u_2 + u_3 = 1$. Let $a = u_1, b = u_2, c = u_3$, we can have $\sum\limits_{i+j+k=n} B^n_{i,j,k}(P) = 1$.

Use the any straight line parallel to one side of the triangle to equate the remaining two sides of the coordinate triangle T into n segments, then the three parallel lines will divide the triangle into n^2 small congruent triangles, thus we can make the n-time subdivision of the coordinate triangle T, recorded as $S_n(T)$. Subsequently, we call each small triangle is the sub-triangle of $S_n(T)$. The vertices of the sub-triangle $((n+1)(n+2)/2$ in all) are called as the node that subdivides $S_n(T)$. The coordinates of the sub-nodes are as follows:

$$(\frac{i}{n}, \frac{j}{n}, \frac{k}{n}), i + j + k = n \tag{4}$$

Abbreviated as:

$$P_{i,j,k} = (\frac{i}{n}, \frac{j}{n}, \frac{k}{n}) \tag{5}$$

Definition 3. *Suppose $b_{i,j,k}(i + j + k = n)$ is any real number, we call*

$$B^n(P) = B^n(u_1, u_2, u_3) = \sum\limits_{i+j+k=n} b_{i,j,k} B^n_{i,j,k}(P) \tag{6}$$

As the n-time Bézier facet on the coordinate triangle T, $b_{i,j,k}$ $(i + j + k = n)$ as the Bernstein coefficient of the Bézier triangular surface, $P_{i,j,k} = (P_{i,j,k}; b_{i,j,k})$, $i + j + k = n$ as the control vertice of the Bézier triangular surface. We call the patch linear continuous function, which is linear on the sub-triangle of $S_n(T)$ and is the value $b_{i,j,k}$ at node $P_{i,j,k}$ as the control grid of the Bézier triangular surface.

In particular, for any function $f : T \to R$, the Bernstein coefficient is taken as:

$$b_{i,j,k} = f(\frac{i}{n}, \frac{j}{n}, \frac{k}{n}) \tag{7}$$

Then we call

$$B^n(P) = B^n(u_1, u_2, u_3) = \sum_{i+j+k=n} f(\frac{i}{n}, \frac{j}{n}, \frac{k}{n}) B^n_{i,j,k}(P) \tag{8}$$

as the n-time Bernstein triangular polynomial of f on T.

Where, in order to simplify the derivation process, three shift operators E_1, E_2, E_3 are introduced, which are defined as:

$$E_1 b_{i,j,k} = b_{i+1,j,k} \tag{9}$$

$$E_2 b_{i,j,k} = b_{i,j+1,k} \tag{10}$$

$$E_3 b_{i,j,k} = b_{i,j,k+1} \tag{11}$$

Subsequently, $E_1^i E_2^j E_3^k b_{0,0,0} = b_{i,j,k}$, and we have:

$$B^n(P) = \sum_{i+j+k=n} \frac{n!}{i!j!k!} u_1^i u_2^j u_3^k (E_1^i E_2^j E_3^k b_{0,0,0}) \tag{12}$$

With the trinomial expansion, Equation (12) can be expressed as:

$$B^n(P) = (u_1 E_1 + u_2 E_2 + u_3 E_3)^n b_{0,0,0} \tag{13}$$

Accordingly, we have:

$$B^n(T_1) = E_1^n b_{0,0,0} = b_{n,0,0} \tag{14}$$

$$B^n(T_2) = E_2^n b_{0,0,0} = b_{0,n,0} \tag{15}$$

$$B^n(T_3) = E_3^n b_{0,0,0} = b_{0,0,n} \tag{16}$$

Here, we call point $P_{n,0,0} = (1,0,0; b_{n,0,0})$, $P_{0,n,0} = (0,1,0; b_{0,n,0})$, $P_{0,0,n} = (0,0,1; b_{0,0,n})$ the corner points of the triangular surface.

When $u_1 = 0$, $u_3 = 1 - u_2$. Substituting into Equation (6), then we have:

$$B^n_{i,j,k}(P) = \frac{n!}{j!(n-j)!} u_2^j (1 - u_2)^{n-j} = B^n_j(u_2) \tag{17}$$

Subsequently,

$$B^n(0, u_2, 1 - u_2) = \sum_{j=0}^n b_{0,j,n-j} B^n_j(u_2), 0 \le u_2 \le 1 \tag{18}$$

The boundary of a triangular surface is the n-time Bézier curve with the boundary of triangular surface control grid as the control polygons.

According to the definition of Bézier triangular surface modeling, when $n = 2$, the quadratic Bézier triangular surface generated by six control vertices is:

$$\begin{aligned}
B^2(P) &= \sum_{i+j+k=2} b_{i,j,k} \frac{2!}{i!j!k!} u_1^i u_2^j u_3^k \\
&= u_1^2 b_{200} + u_2^2 b_{020} + u_3^2 b_{002} + 2u_1 u_2 b_{110} + 2u_1 u_3 b_{101} + 2u_2 u_3 b_{011}
\end{aligned} \tag{19}$$

The above Equation (19) can be further expressed as a quadratic form:

$$B^2(P) = \begin{pmatrix} u_1 & u_2 & u_3 \end{pmatrix} \begin{pmatrix} b_{200} & b_{110} & b_{101} \\ b_{110} & b_{020} & b_{011} \\ b_{101} & b_{011} & b_{002} \end{pmatrix} \begin{pmatrix} u_1 \\ u_2 \\ u_3 \end{pmatrix} \tag{20}$$

The resulting quadratic Bézier triangular surface and its control network projection are shown in Figure 2.

When $n = 3$, the cubic Bézier triangular surface that was generated by ten control vertices is:

$$
\begin{aligned}
B^3(P) &= \sum_{i+j+k=3} b_{i,j,k} \frac{3!}{i!j!k!} u_1^i u_2^j u_3^k \\
&= u_1^3 b_{300} + u_2^3 b_{030} + u_3^3 b_{003} + 3u_1^2 u_2 b_{210} + 3u_1 u_2^2 b_{120} + 3u_1^2 u_3 b_{201} \\
&\quad + 3u_1 u_3^2 b_{102} + 3u_2 u_3^2 b_{012} + 3u_2^2 u_3 b_{021} + 6u_1 u_2 u_3 b_{111}
\end{aligned}
\tag{21}
$$

The resulting cubic Bézier triangular surface and its control network projection are shown in Figure 3.

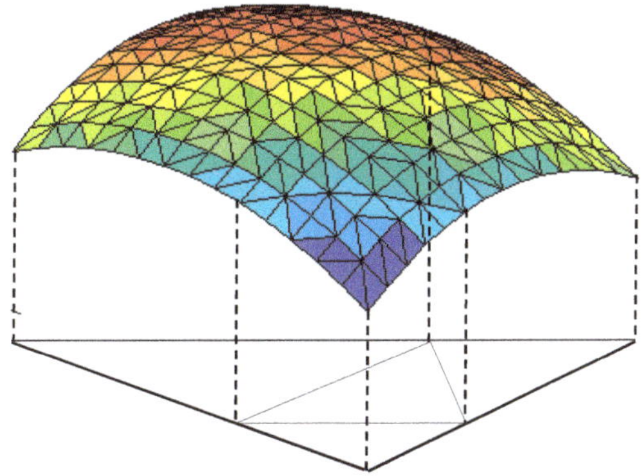

Figure 2. Quadratic Bézier Triangular Surface and Its Control Network Projection.

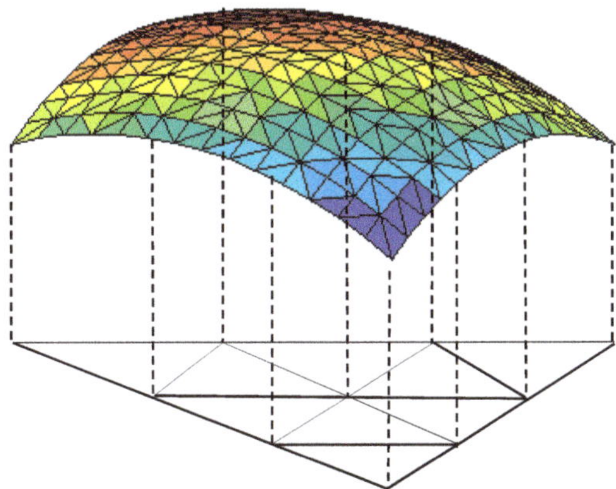

Figure 3. Cubic Bézier triangular surface and its control network projection.

3. Optimal Initial Trajectory Selection for Automatic Spraying on Bézier Surface

3.1. Bézier Surface Definition and Correlative Properties

In particular, the Bézier surface mainly includes a tensor product surface on a rectangular domain and a triangular surface on a triangular domain.

$m \times n$-time Bézier surface can be represented as:

$$B(u,v) = \sum_{i=0}^{m} \sum_{j=0}^{n} B_{i,m}(u) B_{j,n}(v) V_{i,j} \tag{22}$$

where, $B_{i,m}(u)$, $B_{j,n}(v)$ are the u-direction m-time and v-direction n-time Bernstein basis function. $V_{i,j}(i = 0,1,\cdots,m; j = 0,1,\cdots,n)$ is the control vertice or Bézier point of the curved surface. The control vertices form $m+1$ and $n+1$ control polygons along the v-direction and u-direction, respectively, which together form a curved control grid or a Bézier grid.

The properties of the Bézier surface are as follows:

(1) The four corner points of the Bézier grid are the four corner points of the Bézier surface:

$$B(0,0) = V_{0,0}, B(1,0) = V_{m,0}, B(0,1) = V_{0,n}, B(1,1) = V_{m,n} \tag{23}$$

(2) The outermost vertex of Bézier grid defines the four borders of the Bézier surface and it has the following characteristics at the boundary, as shown in Table 1:

Table 1. Characteristics at the Boundary of Bézier Grids.

	(0,0)	(1,0)	(0,1)	(1,1)
B	$V_{0,0}$	$V_{m,0}$	$V_{0,n}$	$V_{m,n}$
$\frac{\partial B}{\partial u}$	$m\Delta^{1,0} V_{0,0}$	$m\Delta^{1,0} V_{n-1,0}$	$m\Delta^{1,0} V_{0,n}$	$m\Delta^{1,0} V_{m-1,n}$
$\frac{\partial B}{\partial v}$	$n\Delta^{0,1} V_{0,0}$	$n\Delta^{0,1} V_{m,0}$	$n\Delta^{0,1} V_{0,n-1}$	$n\Delta^{0,1} V_{m,n-1}$

(3) Affine invariance: The Bézier surface is not changed under affine transformation.
(4) 'Symmetry': The control vertices in opposite order define the same Bézier surface.
(5) Convex hull: The Bézier surface is always located in the three-dimensional convex hull generated by its control vertex.
(6) Move vertice $V_{i,j}$, it will have the largest effect on the point $B(i/m, j/n)$, corresponding to $u = i/m, v = j/n$.

3.2. Optimal Initial Trajectory Selection

A large number of spray painting practical applications show that in the beginning of the spray painting operation if we can determine a good initial trajectory of spray painting robot, the uniformity of paint thickness can be significantly improved. That is to say, the spray painting effect can be improved. It can also reduce the spray painting time, improve the spray painting efficiency, and reduce the rate of paint waste at the same time. At present, the optimal initial trajectory selection method of the existing painting robot is to use the plane cutting method to take the obtained cross line as the initial trajectory of the painting robot [9–11]. This method can improve the spray painting effect to a certain extent, but the randomness is large and the spray painting time cannot be optimized. In this paper, the initial trajectory selection method that is based on geodesic curvature can not only improve the spray painting effect (paint thickness uniformity), but also can improve the spray painting efficiency (reduce spray painting time).

The curvature of a certain point on the painting path can be divided into two kinds: The first one is used to characterize the bending degree of the painting path passing through the point along normal vector of the surface, which is called the normal vector curvature. The second is used to characterize the extent to which the spray trajectory bends to the boundary line of the surface, called geodesic curvature. As shown in Figure 4, Figure 4a is a zero-geometric curvature trajectory. That is, the paint thickness is uniform on the both side of the painting path. Figure 4b is the spray painting trajectory whose geodesic curvature is a constant. It is obvious that more paint is accumulated in the direction where the trajectory is bumped. Figure 4c shows the painting path of the variable geodesic curvature. The paint thickness on both sides of the trajectory is not uniform. It can be seen that the changing rate of the geodesic curvature at each point on the painting path has an influence on the paint thickness. In order to improve the consistency of the paint thickness, the changes in geodesic curvature at each point on the trajectory must be taken into account when selecting the initial spray painting trajectory.

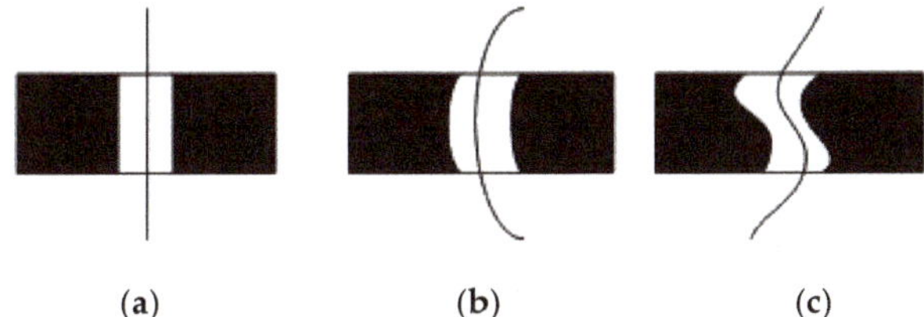

(a)   (b)   (c)

Figure 4. The Influence of Geodesic Curvature on the Consistency of Paint Thickness.

From the example shown in Figure 4, it can be seen that the zero-geometric curvature curve should be selected when determining the initial trajectory, and the initial trajectory also determines the geodesic curvature of the subsequent trajectory. The process of selecting the initial trajectory for spray painting can be divided into two steps: 1) Determining the relative position of the initial spray painting trajectory and the boundary of the workpiece surface. 2) Selecting the direction of the initial spray painting trajectory. As shown in Figure 5a, the initial trajectory is a geodesic, but the geometric curvature is very high when the offset curve passes the area near the apex of the cube. In Figure 5b, the initial trajectory is also the geodesic, but the surface is symmetrically bisected into two parts with the same Gaussian curvature being integral. Thus, in determining the position of the initial trajectory, it is necessary to select a position that minimizes the geodesic curvature of the offset curve.

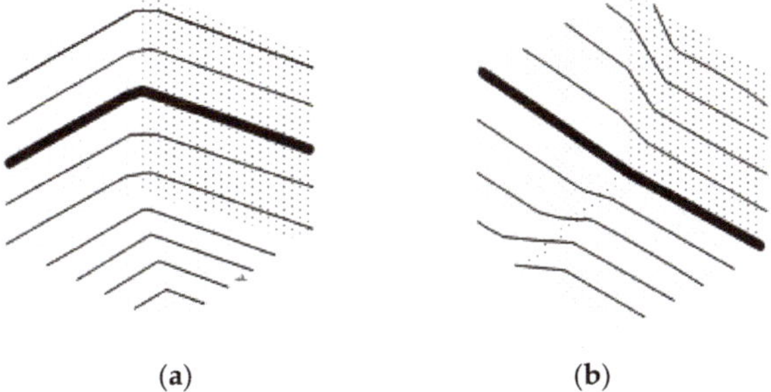

(a)   (b)

Figure 5. The Influence of the Relative Position of Geodesic Curve and Surface Boundary on Subsequent Trajectory.

3.2.1. Determining the Relative Position of Initial Trajectory

As shown in Figure 6, let a segment on the smooth initial trajectory α_0 as C_{st}. We use $\alpha_0(t_0)$ and $\alpha_0(t_1)$ to represent the two endpoints of C_{st}. The offset curve of line segment C_{st} is generated by measuring the distance between the offset curve and the initial trajectory according to the geodesic lines γ_{t_0} and γ_{t_1}, which are perpendicular to the initial trajectory at points $\alpha_0(t_0)$ and $\alpha_0(t_1)$. Assume that C_{of} is the offset curve with a distance Δ of C_{st} and the offset distance Δ is less than the focal length of α_0. Here, we only need to consider the integration of the geodesic curvature along C_{of}. Accordingly, we can assume that the surface is continuous. Subsequently, we can assume that C_{of}, γ_{t_0} and γ_{t_1} are all smooth curves. Assume that ϕ is a region enclosed by boundaries C_{st}, C_{of}, γ_{t_0}, and γ_{t_1}, and an arbitrary smooth curve connecting $\gamma_{t_0}(\Delta)$ and $\gamma_{t_1}(0)$ is C_{dia}. Suppose that the region bounded by C_{st}, γ_{t_0}, and C_{dia} is ϕ_1, its boundary $\partial\phi_1$ contains the curves C_{st}, γ_{t_0}, C_{dia}, and their corresponding directions. Similarly, ϕ_2 denotes the region bounded by C_{of}, γ_{t_1}, and C_{dia}, and $\partial\phi_2$ is the boundary of ϕ_2.

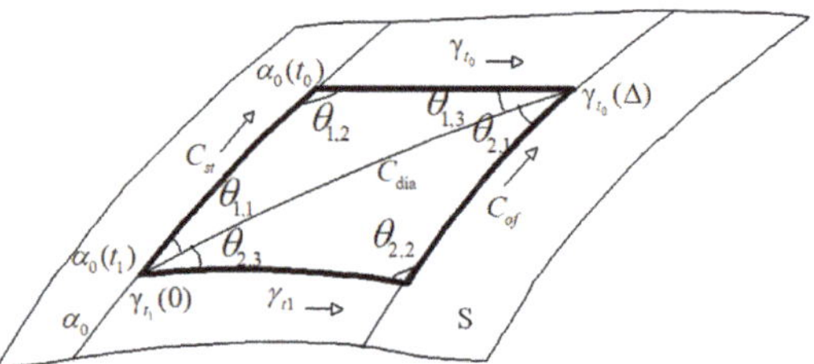

Figure 6. The Relationship between the Curvature Integral of Geodesic Surface and the Gaussian Curvature of Surfaces.

By applying the Gauss-Bonnet formula to the triangular regions ϕ_1 and ϕ_2, we can obtain that:

$$\iint_{\phi_i} K d\sigma + \oint_{\partial\phi_i} k_g ds = \sum_{j=1}^{3} \theta_{i,j} - \pi, \ i = 1,2 \tag{24}$$

In the expression above, K is the Gaussian curvature of the surface ϕ_i. k_g is the geometric curvature of the triangle boundary $\partial\phi_i$. $\theta_{i,j}$ is the j-th interior angle of the boundary $\partial\phi_i$. As γ_{t_0} and γ_{t_1} are geodesic, so the integrals $\int_{\gamma_{t_0}} k_g ds$ and $\int_{\gamma_{t_1}} k_g ds$ are zero. Subsequently, we have:

$$\oint_{\partial\phi_1} k_g ds + \oint_{\partial\phi_2} k_g ds =$$

$$\left(\int_{C_{st}} k_g ds - \int_{\gamma_{t_0}} k_g ds - \int_{C_{dia}} k_g ds \right) + \left(\int_{\gamma_{t_1}} k_g ds + \int_{C_{dia}} k_g ds - \int_{C_{of}} k_g ds \right) \tag{25}$$

$$\oint_{\partial\phi_1} k_g ds + \oint_{\partial\phi_2} k_g ds = \int_{C_{st}} k_g ds - \int_{C_{of}} k_g ds \tag{26}$$

Obviously, $\iint_{\phi_1} K d\sigma + \iint_{\phi_2} K d\sigma = \iint_{\phi} K d\sigma$, γ_{t_0}, and γ_{t_1} are perpendicular to the seed curve, so we have: $\theta_{1,1} + \theta_{2,3} = \theta_{1,2} = \frac{\pi}{2}$, Substituting into expression (26), then we sum the triangular regions ϕ_1 and ϕ_2:

$$\int_{C_{of}} k_g ds = \iint_{\phi} K d\sigma + \int_{C_{st}} k_g ds \tag{27}$$

Finally, if the initial trajectory is a geodesic, then:

$$\int_{C_{of}} k_g ds = \iint_{\phi} K d\sigma \tag{28}$$

3.2.2. Selecting the Direction of Initial Path

In order to ensure that the time along the initial trajectory is the least, when selecting the spatial direction of the initial trajectory, it is necessary to select a curve from numerous geodesic curvature and Gaussian curvature curve on the surface whose integrals are equal to that of an initial trajectory. We can give the definition of surface height first, and then determine the direction of the initial trajectory according to the definition. The surface height is the sum of the longest geodetic curve (straight geodesic) perpendicular to the initial trajectory that extends from the initial trajectory to both sides, as shown in Figure 7. The optimal initial trajectory is the initial trajectory corresponding to the minimum surface height of the surface. With the same spray painting distance, the reciprocating spray painting time along the optimal initial trajectory is the least (Figure 8). What is more, the paint consistency is the best and the paint consumption is also the least.

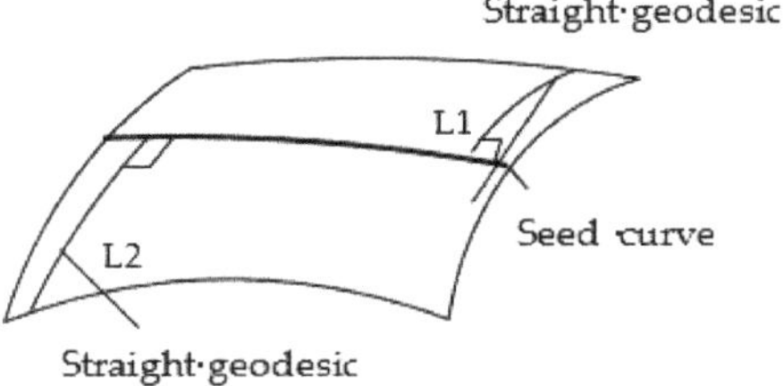

Figure 7. Surface Height Measurement Method.

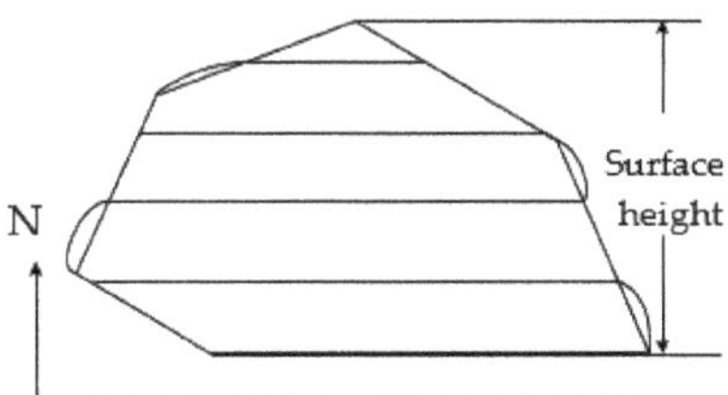

Figure 8. Relationship between Surface Height and Reciprocating Spray Time.

4. Automatic Spray Space Path Generation on Bézier Surfaces

Under normal circumstances, using the Bézier method to generate the surface is more complex. It is more complex if we perform optimization of spraying trajectory directly on Bézier surface. On the other hand, during the spraying operation, the distance between the end effector and the workpiece surface is always constant and perpendicular to the workpiece surface. In this case, the end effector of the spray equipment is essentially reciprocating on the isometric surface of the workpiece surface. Therefore, according to this idea, we can first find its isometric surface according to the shape of the surface and then perform optimization of spraying trajectory on the isometric surface. It should be noted that, strictly speaking, it is the discrete point array of the isometric surface of the Bézier surface but not the isometric surface of Bézier surface that we will find out according to this method. In the spatial path planning of the spray painting robot, we only need to find the discrete points on the path essentially and then fit out the spray path using the corresponding mathematical methods according to the accuracy requirements.

A U-direction or V-direction Bézier curve on the Bézier surface boundary is used as a benchmark and it is discretized under a certain precision. Subsequently, specify a constant painting distance h and find the equidistant point of the discrete points on the Bézier surface in the direction of the normal vectors of the discrete points along the curve. After connecting these equidistant points with a smooth curve, we can find an equidistant line of a Bézier curve on the boundary line of the Bézier surface. By the same token, the Bézier curves with the same distance are specified on the Bézier surface (which is the distance between two adjacent painting paths). In the same way, the same method can be used to find the discrete point array of the equidistant surfaces of the Bézier surface. Afterwards, we use the cubic Cardinal spline curves to connect each discrete point array. The adjacent two segments of cubic Cardinal spline curve segments are connected by a Hermite spline curve, so that the specified painting path can be obtained.

5. Trajectory Optimization on the Bézier Surface

In the actual off-line programming process of spray painting robot, the following factors should be taken into consideration when performing the trajectory optimization for spray painting robot on curved surface: (1) Mathematical model of surfaces. (2) Spray painting model on curved surface. (3) The expression of paint thickness at a point on the surface. (4) Mathematical expression of optimized trajectory on surface and its solution. In essence, the trajectory optimization for spray painting robot is actually a multi-objective optimization problem with constraints. There are many constraints in this problem, such as the error of paint thickness, the path of the end effector, the moving speed, the surface shape of the sprayed workpiece, parameters of the end effector, air pressure, paint viscosity, and so on. Accordingly, how to deal with the constraint function effectively in order to guide the algorithm searching is the key of trajectory optimization problem [12–15]. On the other hand, there are many optimization objectives, such as minimum spray painting time, smallest variance of paint thickness, minimum paint consumption, the highest paint utilization rate, the least inflexion of the robot path, and so on. In these spray painting optimization objectives, the objective function of the trajectory optimization is not independent of each other. They are often coupled with each other and in a state of competition. As a result, it is very difficult to obtain the exact solution of the multi-objective trajectory optimization problem of spray painting robot.

In order to obtain high-efficient painting path, the ideal method is to make certain assumptions. That is, in the case that the error is allowed, a number of parameters are assumed to remain unchanged in the process of spray painting. Only the main factors in the spray painting process are taken into account. Such kind of idea makes the trajectory optimization of spray painting robot greatly simplified and it also makes the multi-objective optimization of spray painting trajectory with constraints being easy to be solved.

When solving the optimization problem of spray painting trajectory on Bézier surfaces, we will simplify and solve the problem according to the ideal above. The specific idea is as follows: After the Bézier triangular surface modeling method is used to obtain the parametric surface, a simple paint deposition rate model is established. On this basis, a general spray painting model on the Bézier surface is derived and the mathematical expression of paint thickness at an arbitrary point is also derived. Finally, the optimal spray painting speed and spray painting time are selected as optimization objectives. After the multi-objective optimization function of the spray painting robot on Bézier surface is derived, the appropriate mathematical programming method is used to obtain the solution and the optimized trajectory of spray painting robot on Bézier surface can also be obtained.

The spatial distribution model of coating, the cumulative rate of coating function diagram and free surface trajectory optimization method have been described in the previous work [3,16]. After spraying a curved surface S, assuming that the average thickness of the surface is q_d, the coating thickness at any point s is q_s, the deviation of the maximum coating thickness is q_w, and then we have:

$$\max_{s \in S}(|q_d - q_s|) \leq q_w \tag{29}$$

Assuming that the maximum coating thickness is q_{max} and the minimum coating thickness is q_{min}, the maximum deviation angle between the normal vector of all sampling points and the normal vector of the surface is β_{th}, then the coating thickness at any point s can satisfy the following inequality:

$$q_{min} \cdot \cos \beta_{th} \leq q_s \leq q_{max} \tag{30}$$

The coating thickness at any point s satisfies the requirement (29), then we have:

$$|q_s - q_d| \leq q_w, s \in S \tag{31}$$

then:

$$q_{max} - q_d \leq q_w \tag{32}$$

Further:

$$q_d - q_{min} \cdot \cos \beta_{th} \leq q_w \tag{33}$$

If Equation (32) can be satisfied, then the maximum deviation angle β_{th} can be calculated with Equation (33). That is, for any surface, if the deviation angle β satisfies $\beta \leq \beta_{th}$, then the coating thickness at any point on the curved surface can satisfy Equation (29).

6. Experimental Part

6.1. Experimental Verification

The sprayed workpiece is shown in Figure 9. According to the topology of the spray workpiece, the workpiece is divided into three parts for processing, which are basin bottom, basin side, and basin edge, respectively. In the three parts, the basin bottom and the basin edge are all flat, and the surface of these parts can be directly generated by the control vertice. The basin side needs to be divided into two patches for processing, which are both arc. Where 10 control vertices are taken at different positions on each patch, and the cubic Bézier triangular surfaces are generated using the algorithm in Part 2. The Bézier triangular surface generation software system is written in VC ++ language. The curved surface modeling diagram of the workpiece is shown in Figure 10.

After the sprayed workpiece is modeled by the Bézier triangular surface technique, the U-direction spatial path and the V-direction spatial path of the workpiece surface are obtained according to the method for generating the spatial path of the spray painting robot on the Bézier surface that is presented in this paper. U-direction Spatial Path and V-direction Spatial Path as shown in Figures 11 and 12.

Figure 9. Sprayed Workpiece.

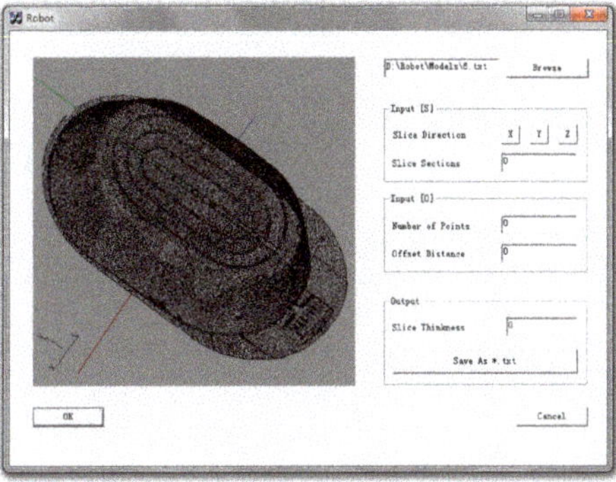

Figure 10. The Curved Surface Modeling Diagram of the Workpiece.

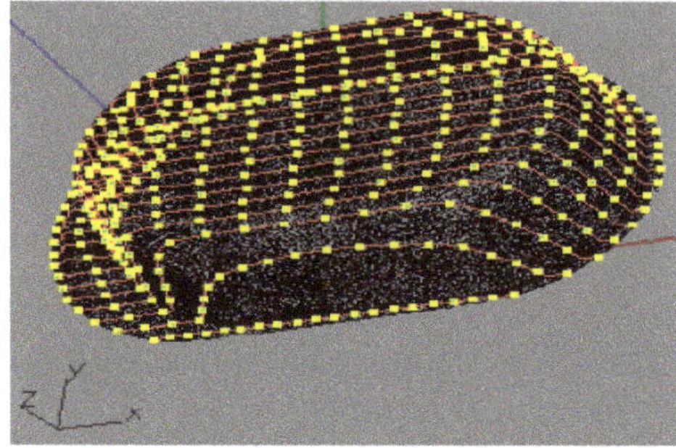

Figure 11. U-direction Spatial Path.

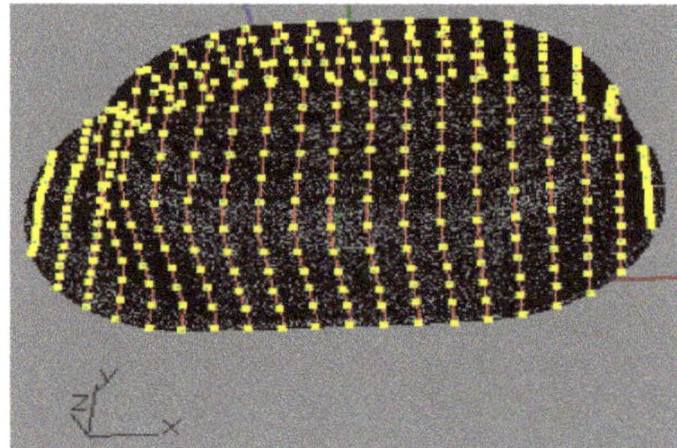

Figure 12. V-direction Spatial Path.

Assuming that the ideal paint thickness is $qd = 50$ μm, the error of the maximum paint thickness is $qw = 10$ μm, the bottom radius of the conical paint sprayed by the end effector R = 60 mm. According to the Spatial distribution model of coating, after performing the spray painting experiment on the plate, the paint deposition rate obtained by the experimental data is:

$$f(r) = \frac{1}{15}(R^2 - r^2) \ \mu m/s \tag{34}$$

After obtaining the optimized trajectory on the plane, the spray painting rate (at uniform speed) of the spray painting robot can be obtained as: V = 256 mm/s.

After obtaining the spatial painting path, trajectory optimization is carried out along the specified painting path according to the trajectory optimization method for spray painting robot on the Bézier surface in Section 5. At the same time, according to the initial trajectory selection method for spray painting robot on the Bézier surface in Section 3, the initial trajectory is selected. There are 432 discrete points in the discrete point array in U-direction, and the path between every two discrete points is divided into 10 segments. There are 402 discrete points in the discrete point array in the V-direction, and the path between every two discrete points is divided into 10 segments. The parameters of the algorithm are as follows: the ideal paint thickness $qd = 50$ μm, maximum allowable error $qw = 10$ μm, painting radius R = 50 mm, painting distance h = 100 mm, numbers of triangular facets N = 1566, the length of each segment $dk = 50$ mm, number of the subdivided segments m = 10, and weight vector $\omega = (0.5, 0.5)^T$. When the gun paints at uniform speed, v = 256 mm/s (the optimization speed on the plane). Take v = 256 mm/s as the initial value of the algorithm iteration when performing optimized spray painting. The following optimization experiments are carried out in the U-direction path and the V-direction path, respectively. The process of spray painting experiment in the laboratory is shown in Figure 13. After the spray painting experiment, the paint thickness is measured by a paint thickness gauge. The paint thickness curve of the 432 sampling points along the U-direction path is shown in

Figure 14 and the paint thickness curve of the 402 sampling points along the V-direction path is shown in Figure 15. The experimental results are shown in Table 2.

After the analysis of the experimental results, we can learn that spray painting along the U-direction and the V-direction can both meet the spray painting requirement after the optimization of spray painting trajectory. That is, the error of paint thickness is within the allowable range. However, it can be seen that the spray painting effect along the U-direction path is better and the spray painting efficiency is higher for the workpiece. It also can be seen that the shape of the workpiece surface should be fully considered in the planning of painting path, and the direction of the painting path may have a certain impact on the spray painting effect and efficiency.

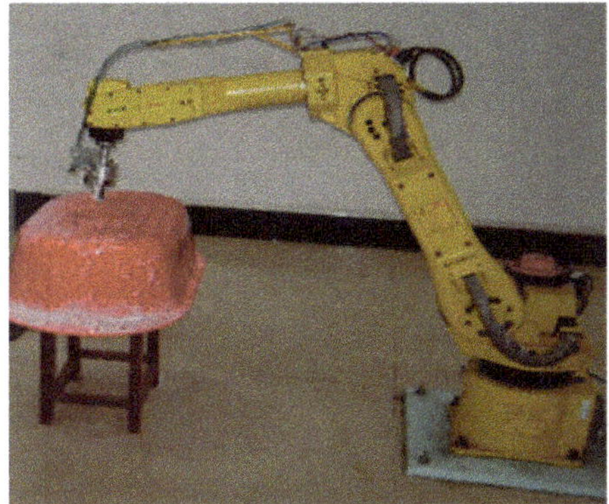

Figure 13. Spray painting Experiment in the Laboratory.

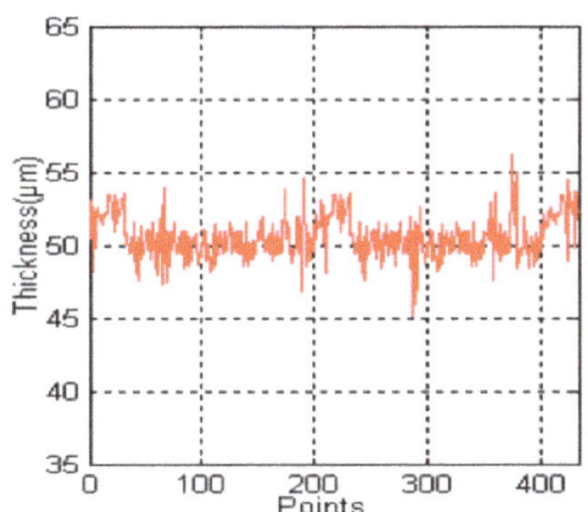

Figure 14. The Paint thickness Curve Spray painting along the U-direction Path.

Table 2. The Experimental Results of Spraying.

	Optimized Spray Painting along U-Direction Path	Optimized Spray Painting along V-Direction Path
Average (μm)	51.2	52.1
Maximum (μm)	56.2	58.3
Minimum (μm)	45.1	43.0
Painting time (s)	83	95

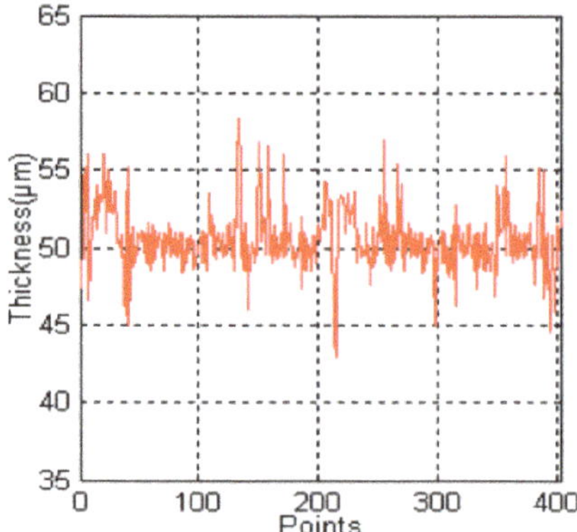

Figure 15. The Paint thickness Curve Spray painting along the V-direction Path.

6.2. Spray Painting Experiment

Taking a certain brand of automobile body as the spraying target, the feasibility of automatic spraying trajectory optimization on the Bézier surface was simulated. As shown in Figure 16, taking the automotive body of a brand as the paint objective and taking the U direction as the spraying direction. Four ABB robots are using for painting at the same time. After the painting is completed, the coating thickness of the sample points on the surface of the car body is measured by the paint drying and the professional coating thickness tester.

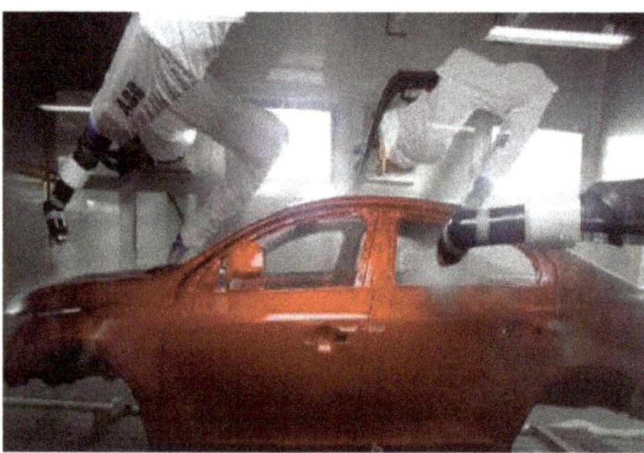

Figure 16. The robotic spray painting experiment.

In the spray experiment, the ideal paint thickness is qd = 50 μm, maximum allowable error qw = 10 μm, painting radius R = 50 mm, painting distance h = 100 mm, and painting speed V = 256 mm/s (the optimization speed on the plane) when performing uniform spray painting. We take 400 discrete points evenly on the workpiece surface after the spray painting operation. The paint thickness curve is shown in Figure 17 after using a paint thickness gauge to measure the paint thickness at the discrete points.

Based on the analysis of the experimental results, it can be seen that the average spray thickness is 51.1 μm, the thickness of the maximum coating is 58.1 μm, the minimum coating thickness is 44.2 μm, and the spray time spent by the robot is about 99 s, which is better than the general spraying robot. After the spraying trajectory is normalized based on the Bézier surface, the spraying requirements can be met along the path spraying, that is, the coating thickness deviation is within the allowable range.

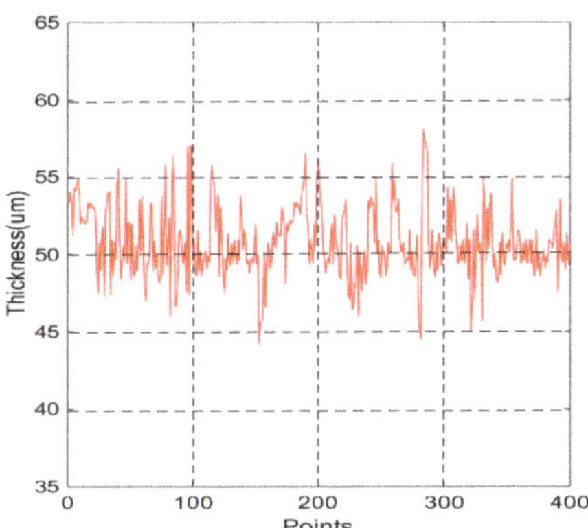

Figure 17. Material Thickness of Random Chosen Points for U-direction Trajectory.

7. Conclusions

In this paper, an automatic spray trajectory optimization method on Bezier surface is proposed. Firstly, the Bézier surface is analyzed and the method for searching the optimal initial trajectory of the spray painting robot is given according to the features of the Bézier triangular surface. Subsequently, the spray painting model of Bézier surface is established, and the mathematic expression of paint thickness at a certain point on Bézier surface is given. Finally, the optimized trajectory of Bézier surface is obtained by using the ideal point method in the mathematical programming with the uniformity of paint thickness and the shortest spray painting time as optimization objectives along the specified spray painting path. The biggest advantage of this method is that a good initial path of automatic spraying is determined at the beginning of the spraying process, which can significantly provide uniformity of coating thickness, that is, improve the spraying effect. Finally, the effectiveness and practicability of the proposed method are verified by an example verification and spraying experiment.

Author Contributions: W.C. and Y.T. conceived and designed the experiments; W.C. performed the experiments; H.G. analyzed the data; J.L. contributed reagents/materials/analysis tools; J.L. wrote the paper.

Funding: This research is supported by the Postgraduate Research & Practice Innovation Program of Jiangsu Province (KYCX18_2332), National Natural Science Foundation of China (51505193, 61503162), Project funded by China Postdoctoral Science Foundation (2016M601691), Industrial Foresight Project of Zhenjiang City (GY2018018) and The six highest peak of talent in Jiangsu Province (2016-GDZB-021).

Conflicts of Interest: The authors declare no conflict of interest.

Annotation

T_1, T_2, T_3	Clockwise vertex of any given triangle.
P	Any point in the plane where the triangle $T_1 T_2 T_3$ is located.
$[T_1 T_2 T_3]$	Represents the directed area of the triangle $T_1 T_2 T_3$.
T	The interior of the triangle $T_1 T_2 T_3$ with the boundary.
$S_n(T)$	Make the n-time subdivision of the coordinate triangle T.
$B_{i,m}(u), B_{j,n}(v)$	The u-direction m-time and v-direction n-time Bernstein basis function.
$B(0,0), B(1,0), B(0,1), B(1,1)$	The four corner points of the Bézier grid.
$V_{0,0}, V_{m,0}, V_{0,n}, V_{m,n}$	The four corner points of the Bézier surface.
α_0	The smooth initial trajectory.
C_{st}	A segment on the smooth initial trajectory.
$\alpha_0(t_0), \alpha_0(t_1)$	The two endpoints of C_{st}.
$\gamma_{t_0}, \gamma_{t_1}$	The geodesic lines.
C_{st}	The offset curve of line segment.
Δ	The offset distance.
C_{of}	The offset curve with a distance Δ of C_{st}

ϕ	A region enclosed by boundaries C_{st}, C_{of}, γ_{t_0} and γ_{t_1}
C_{dia}	Any smooth curve connecting $\gamma_{t_0}(\Delta)$ and $\gamma_{t_1}(0)$
ϕ_1	The region bounded by C_{st}, γ_{t_0} and C_{dia}
ϕ_2	The region bounded by C_{of}, γ_{t_1} and C_{dia}
$\partial\phi_1$	The boundary of ϕ_1
$\partial\phi_2$	The boundary of ϕ_2

References

1. Li, J. Distributed cooperative tracking of multi-agent systems with actuator faults. *Trans. Inst. Meas. Control* **2015**, *37*, 1041–1048. [CrossRef]
2. Li, J.; Li, C.; Yang, X.; Chen, W. Event-triggered containment control of multi-agent systems with high-order dynamics and input delay. *Electronics* **2018**, *7*, 343. [CrossRef]
3. Chen, W.; Liu, J.; Tang, Y.; Huan, J.; Liu, H. Trajectory optimization of spray painting robot for complex curved surface based on exponential mean Bézier method. *Math. Probl. Eng.* **2017**, *2017*, 4259869. [CrossRef]
4. Li, F. Trajectory optimization of spray painting robot based on CAD. *Trans. Chin. Soc. Agric. Mach.* **2010**, *41*, 213–217.
5. Gong, D. Trajectory optimization of spray painting robot for natural quadric surfaces. *China Mech. Eng.* **2011**, *22*, 282–290.
6. Huang, W.H. Optimal line-sweep-based decompositions for coverage algorithms. In Proceedings of the IEEE International Conference on Robotics and Automation, Seoul, South Korea, 21–26 May 2001; Volume 1, pp. 27–32.
7. Kim, T.; Sarma, S.E. Optimal sweeping paths on a 2-manifold: A new class of optimization problems defined by path structures. *IEEE Trans. Robot. Autom.* **2003**, *19*, 613–636.
8. Gasparetto, A.; Vidoni, R.; Pillan, D.; Saccavini, E. Automatic path and trajectory planning for robotic spray painting. In Proceedings of the 7th German Conference on Robotics, Munich, Germany, 21–22 May 2012; pp. 1–6.
9. Zhao, J. The Influence of initial configura tion on trajectory planning algorithms for redundant robots. *Mech. Technol.* **2000**, *19*, 249–251.
10. Chen, W.; Tang, Y.; Zhao, Q. A novel trajectory planning scheme for spray painting robot with Bézier curves. In Proceedings of the Control and Decision Conference, Yinchuan, China, 28–30 May 2016; pp. 6746–6750.
11. Chen, H.; Fuhlbrigge, T.; Li, X. A review of CAD-based robot path planning for spray painting. *Microelectron. Reliab.* **2009**, *42*, 343–347. [CrossRef]
12. Posa, M.; Cantu, C.; Tedrake, R. A direct method for trajectory optimization of rigid bodies through contact. *Int. J. Robot. Res.* **2013**, *33*, 69–81. [CrossRef]
13. Shao, Z.J.; Shao, Z. *Precise Trajectory Optimization for Articulated Wheeled Vehicles in Cluttered Environments*; Elsevier Science Ltd.: Amsterdam, The Netherlands, 2016.
14. Tang, Y.; Chen, W. Surface modeling of workpiece and tool trajectory planning for spray painting robot. *PLoS ONE* **2015**, *15*, e0127139.
15. Liang, X.; Wan, L.; Blake, J.I.R.; Shenoi, R.A.; Townsend, N. Path following of an underactuated AUV based on fuzzy backstepping sliding mode control. *Int. J. Adv. Robot. Syst.* **2016**, *13*, 1. [CrossRef]
16. Chen, W.; Liu, H.; Tang, Y.; Liu, J. Trajectory optimization of electrostatic spray painting robots on curved surface. *Coatings* **2017**, *7*, 155. [CrossRef]

 *electronics*

Article

Control System in Open-Source FPGA for a Self-Balancing Robot

Juan Ordóñez Cerezo [1,†], **Encarnación Castillo Morales** [2,†] **and José María Cañas Plaza** [1,*,†]

1   RoboticsLab-URJC, Rey Juan Carlos University, Fuenlabrada, 28943 Madrid, Spain; jordonezcerezo@hotmail.com
2   DiTEC Research LAB, Granada University, 18071 Granada, Spain; encas@ugr.es
*   Correspondence: josemaria.plaza@urjc.es; Tel.: +34-914-888-755
†   These authors contributed equally to this work.

Received: 30 December 2018; Accepted: 1 February 2019; Published: 9 February 2019

Abstract: Computing in technological applications is typically performed with software running on general-purpose microprocessors, such as the Computer Processing Unit (CPU), or specific ones, like the Graphical Processing Unit (GPU). Application-Specific Integrated Circuits (ASICs) are an interesting option when speed and reliability are required, but development costs are usually high. Field-Programmable Gate Arrays (FPGA) combine the flexibility of software with the high-speed operation of hardware, and can keep costs low. The dominant FPGA infrastructure is proprietary, but open tools have greatly improved and are a growing trend, from which robotics can benefit. This paper presents a robotics application that was fully developed using open FPGA tools. An inverted pendulum robot was designed, built, and programmed using open FPGA tools, such as IceStudio and the IceZum Alhambra board, which integrates the iCE40HX4K-TQ144 from Lattice. The perception from an inertial sensor is used in a PD control algorithm that commands two DC motors. All the modules were synthesized in an FPGA as a proof of concept. Its experimental validation shows good behavior and performance.

Keywords: robotics; open FPGAs; robot control

---

1. Introduction

The most common approach taken for the computing required in technological applications is using software which writes instructions for a general-purpose circuit, such as a Computer Processing Unit (CPU) or Graphical Processing Unit (GPU) [1]. Another option is designing a special circuit for this specific computation, where Application-Specific Integrated Circuits (ASICs) [2] are the traditional hardware implementation for system design. This last alternative requires more effort and has high development costs, but when an application requires real-time processing, such as a video, television, or robotic controller for real-time trajectory generation, the requirements are highly demanding and better met when implemented in hardware. However, in general, applications are more flexible when implemented in software rather than in hardware, especially when they are not computationally demanding or when they are non-critical. The emergence of Programmable Logic Devices (PLD) [3] and reconfigurable devices, such as Field-Programmable Gate Arrays (FPGAs) [4] have changed this scenario. The FPGAs are a well-established technology, not only for prototyping and development, but also as a for ASICs in a growing number of applications, as they offer benefits very similar to those of ASICs, such as high speed and reliability, but without requiring as much resources or costs as the custom ASIC design [2]. In addition, these features go far beyond those possible with microprocessor-based systems, while maintaining similar flexibility thanks to their reconfigurability. Unlike microprocessors, FPGAs perform different operations in parallel, and it is unnecessary to compete for the same resources. Thus, incorporation of the FPGA in the industry has been driven by the fact that FPGAs are the combination of the best features of ASICs and microprocessor-based systems. Fields in which FPGAs are currently used include medical imaging, coding and encryption, aeronautics and defense, voice recognition, artificial vision, and robotics.

Thus, FPGAs are definitively established in the digital systems market, with Lattice Semiconductor Corp. [5], Xilinx [6], and Intel FPGA [7] as the main private companies. Xilinx, which recently signed a large collaboration agreement with IBM, and Intel FPGA, the new trade name of Altera after its acquisition by Intel, are market leaders. FPGA devices generally consist of a regular matrix of logic blocks and an interconnection network, both configurable, together with multiple I/Os. The high-end segment [8,9] has also integrated specific resources for digital signal processing, support for networks, or embedded microprocessors, usually being ARM cores. Thus, synthesized embedded microprocessors are included in FPGA devices, such as Nios II [10] or MicroBlaze [11]. More recently, RISC-V [12], a free and open RISC instruction-set type of architecture, has been implemented within Microsemi FPGA [13,14]. Moreover, FPGAs have a strong presence in the sector of artificial intelligence [15] providing hardware accelerators in this field that can exceed the performance of GPUs. Despite these advanced features, it is worth noting that these are proprietary FPGAs, and working with them requires a large budget that is not always feasible, such as in educational applications.

ISE from Xilinx and Quartus II from Intel FPGA are proprietary software tools offered by these companies for synthesis and analysis of designs to be implemented into its FPGAs, usually using a Hardware Description Language (HDL) [16]. These software tools enable the developer to synthesize or compile their designs, to examine RTL descriptions, to perform timing analysis, to simulate the designs, and to configure the target device using the programmer [17]. There are currently two industry standard HDLs: VHDL (very high-speed integrated-circuit Hardware Description Language) [18] and Verilog [19]. To compare these two, on the one hand, VHDL is strongly typed, it has the ability to define custom types, it can define multiple signals into one type, and the logical statement endings are clearly marked. However, it is also extremely verbose, and needs sensitivity lists and type conversions. On the other hand, Verilog is a compact language, performs logical tests on an entire array of bits with a single operator, and is adequate for low-level descriptions closer to the actual hardware. But nevertheless, it is a weakly typed language, it offers no support of custom types, the signal declarations can be confusing, and it has reduced support for asynchronous signals. This scenario has forced the search and development of new alternatives, such as SpinalHDL [20], an open-source high-level called whose goal is to use simple elements (flip-flops, gates, if/case statements) to create a new abstraction level and help the designers to reuse their code. Among their advantages over VHDL and Verilog are the evolving capabilities, the reduction of the code size, the easy type conversions, the loop detection, and that it is free and has a user-friendly IDE.

As mentioned previously, robotics is one of the application fields of FPGA [21–23]. Typically, the implementation of robot intelligence and controllers in FPGAs provides many advantages, like reliability and fast operation, which allow for better robot control. However, a large budget is required to work with proprietary FPGA software tools and it is not affordable for educational applications, such as educational robotics [24]. This paper presents a novel use of open-source FPGAs for educational robotics using a new visual language for robot programming. Concretely, an inverted pendulum real robot was developed. The main characteristics of this implementation and the performed experiments confirm the feasibility of this proposal.

The rest of the manuscript is organized as follows: Section 2 is devoted to the related works involved in open FPGA in robotics and the inverted pendulum, while Section 3 describes the design of the self-balancing robot using open FPGA. Section 4 presents the experiments, where a real implementation of the self-balancing robot over an open FPGA is carried out, confirming the feasibility of the presented proposal. Finally, the main conclusions are presented in Section 5.

2. Related Works

Three areas provide the context for the proposal in this study: the use of FPGAs in robotics, the open FPGA community, and the robotic application selected as a proof of concept—the self-balancing robot. Some key works are also reviewed in this section.

2.1. FPGA in Robotics

Robotics is one of the fastest-growing technological areas in recent years [25]. It is based on systems composed of mechanisms which are able to make movements and execute specific tasks that are programmable and intelligent. Some implementation solutions for digital control systems for robot manipulators and mobile robots proposed in the literature use hardware technologies, such as DSPs or microprocessors [26]. These solutions allow for real-time control, but since the DSP has limited output ports, applications for control of humanoid robots, for instance, are not suitable. FPGA technology avoids this limitation, ultimately reducing size and weight, and therefore, costs. In addition, due to the efficient integration of embedded processors' intellectual properties (IPs) into a FPGA, the highly sophisticated algorithms with heavy computations required by robotic controllers can be performed by software in an FPGA. Thus, many FPGA-based solutions have been implemented in the field of robotics, such as a static gesture recognition system [21], an algorithm for collision detection between Oriented-Bounding-Boxes (OBBs) [22], and an embedded, robust adaptive controller for mobile robotics [23]. Many different works have shown that FPGA implementation of robotic applications is the best solution for optimum performance. Robotics may generate benefits not only in the industry, but also in classrooms [24], enabling the emergence of new learning systems. In addition, in a future world where robots will be used in almost any activity, a learning approach using these systems in the classroom enables students' technological development at an early age, facilitating their integration into the adult world. The following are some of the educational benefits of robotics: they drive initiative and creativity; promote greater sociability; encourage algorithmic and mathematical thinking; facilitate teamwork, problem solving, and active learning; and enhance self-esteem. However, in order to facilitate educational robotics in the classroom, the systems must consider the following elements:

- A high technological integration level is not recommended;
- Robots must be sociable and fun;
- Programming frameworks should not be complex, their functionality should be limited to a certain extent, and they must attract students' attention and make them feel comfortable in the context;
- It is important for the robot to have a series of sensors and actuators, as well as inputs and outputs so that the results are visual.

A major obstacle to achieving these features in educational robotics is that most commercial educational robot platforms are closed. Thus, robot vendors do not commonly provide support for old control units, or when a deprecated robot requires an update or even simple preventive maintenance. The manufacturer tends to recommend disposing of such a unit and acquiring a new one. It is worth mentioning the well-known LEGO, PID control [27], and walking robot [28]. Open FPGAs and their new graphical IDE tools may help to avoid this obstacle, as described in the following section.

2.2. Open FPGAs

Many HDLs, as well as FPGA architectures are linked to important companies, such as Xilinx and Intel FPGA, and working with them entails high development costs. Hence, not many companies or individuals can benefit from the advantages of using FPGAs, meaning FPGA technology progresses at a slower pace. One of the keys to the success of companies like Arduino [29] is the large community of people that stands behind creating new libraries, components, etc. This is mainly due to the low price of its products and the possibility of finding all the hardware and software on the web. To understand the creation of open FPGA [30], it is important to understand the bitstream that is used to describe the configuration with which a specific design will be implemented in a FPGA. This detailed bitstream format for a particular FPGA is typically owned by the FPGA manufacturer. This is why Cliford Wolf decided to interpret the bitstream of the Lattice iCE40 FPGA devices [5] and developed the IceStorm tool [31], a translate software from Verilog and bitstream. This translation was possible thanks to the inverse engineering, meaning that it is not the usual usage that is given, but the inverse. Thus, there is no longer dependency on any manufacturer, and all knowledge is also available. From these tools, new interfaces or new applications that were not foreseen by the manufacturer can be created. Nowadays, the focus of the IceStorm project is on HX1K-TQ144 and HX8K-CT256 devices from Lattice iCE40, but since it is an open project, a lot of people are increasing their chances. IceZum Alhambra [32] is an FPGA development board including the open FPGA iCE40HX4K-TQ144 from Lattice. It is an open hardware that is compatible with the IceStorm toolchain and with Arduino Uno shields. Figure 1 shows the Icezum Alhambra II Board. Important features of this board include the following:

- 12 MHz oscillator;
- On/off switch to enable or disable digital pins;
- 20 Input/Output 5 V pins;
- 8 Input/Output 3.3 V pins;
- Micro-B USB to program FPGA from PC;
- Reset button;
- Eight general-purpose LEDs;
- TX/RX LEDs;
- 4 analog inputs available through i2c;
- 8 K memory;
- Possibility of powering through LIPO battery.

Figure 1. Icezum Alhambra II Board.

This development board can be implemented with new open tools, like IceStudio [33], a graphical IDE for free FPGAs, built on the IceStorm project. It provides simple tools to analyze and create bitstream files—that is, the lowest level of implementation for an FPGA. Boards with better features do exist, but IceZum II Alhambra provides open hardware that can be implemented with free and open software tools. This board was created with the idea of making digital electronics user-friendly for young students, allowing for a visual language for programming the FPGA [34], fulfilling the aforementioned features required for educational robotics.

2.3. Inverted Pendulum

The inverted pendulum is of one the most famous problems in terms of control theory and systems dynamics [35,36]. An inverted pendulum is represented in Figure 2, and consists of a pendulum where the center of mass is located above the balancing axis. Maintaining an upward equilibrium position is a challenge, as this equilibrium position is unstable (a system is more stable when its center of mass is closer to the supporting horizontal plane). As the inverted pendulum system is non-linear, it is well-suited for control by fuzzy logic [37]. The inverted pendulum system has significant theoretical value since it represents the basis of many complex systems, such as biped robot upright walking balance control, rocket launch vertical control, spacecraft attitude control, and offshore drilling platform stability control. Beyond its theoretical interest, the inverted pendulum is also attractive for university professors of engineering and teachers in secondary education. In this paper, a solution for the inverted pendulum problem is addressed using an open FPGA in order to correct its instability.

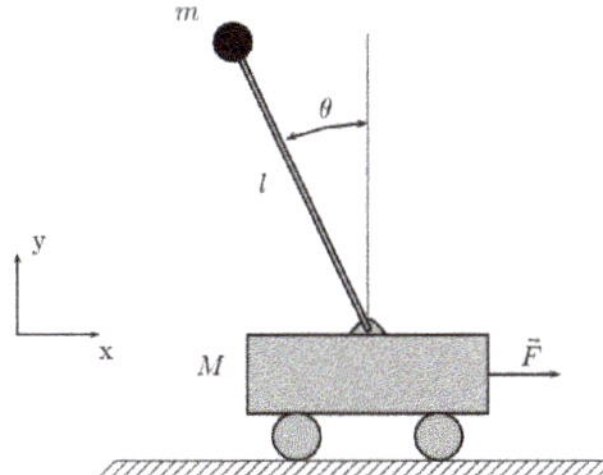

Figure 2. Representation of an inverted pendulum.

3. Self-Balancing Robot

In this section, the proposed solution for the inverted pendulum problem through the use of an FPGA in coexistence with a micro-controller is described. Several aspects are addressed, such as the physics of the self-balancing robot used in the experiments, the calculation of its structure, the sensors and actuators used, the control system, and the design and manufacture of a Printed Circuit Board (PCB) to solve certain engineering problems.

This section begins with a brief high-level description (Section 3.1) and continues with the details of the perception element (Section 3.2). Subsequently, the connection between Arduino and FPGA (Section 3.3), the PD control on FPGA (Section 3.4), and a motor block in IceStudio (Section 3.5) are also described.

3.1. Design

The hardware design of the inverted pendulum control with FPGA is shown in Figure 3. The microcontroller obtains the current angle of the system by means of i2c communication with an Inertial Measurement Unit (IMU) sensor. In the microcontroller, once the current robot vertical angle is read, a serial-type communication sends it to the FPGA in a binary format of 1 byte for the integral part and 1 byte for the decimal part. Inside the FPGA, the robot angle is read and the speed commands to the motors for correction of the angle are calculated by a basic PD controller. A shield with a DC motor driver is connected to the FPGA, and provides the possibility of varying the speed and the movement direction of two DC motors that permit the stabilization of the system.

3.2. Perception

Continuous knowledge of the angle is necessary for its analysis and correction. For this purpose, the MPU6050 sensor was used, and connected to an Arduino Nano by i2c communication. In order to correct some of the data collection problems such as noise or drift, it incorporates an internal processor

(Digital Motion Processor, DMP) that executes data fusion algorithms (Motion Fusion) to combine the measurements of the internal sensors, avoiding the necessity of performing the filters externally (Figure 4).

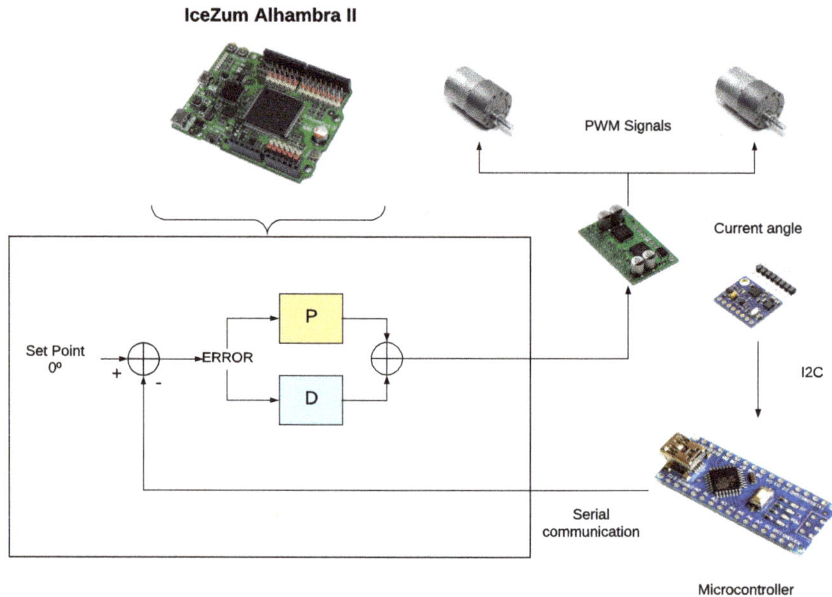

Figure 3. Hardware design of the inverted pendulum control with Field-Programmable Gate Arrays (FPGA).

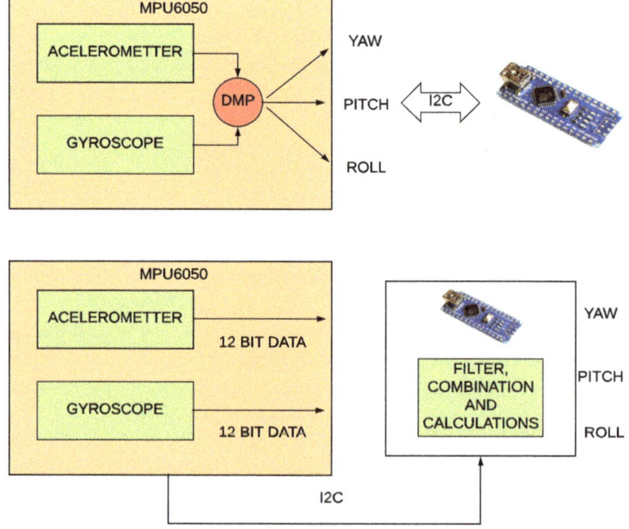

Figure 4. Advantage in the use of DMP.

3.3. Arduino-FPGA Connection

An integration between the microcontroller and the FPGA allows sequential and parallel tasks to be distinguished, assigning each process to the microcontroller if it the task needs to be sequential, or to the FPGA if the process can be parallelized, thus obtaining certain advantages. More than one option exists for the microcontroller/FPGA integration. In this work, physical coexistence with communication between them was chosen. They can also be integrated by means of several types of communication. Serial communication was selected as the most appropriate type due to the numbers of pins available in the FPGA. The communication would only be unidirectional, with the microcontroller sending information to the FPGA about the current angle of the robot in order for the FPGA to analyze and actuate starting from that angle. There are thus two parts in this serial communication: from the point of view of the microcontroller, and from the point of view of the FPGA. The sensor reading acquisition in the microcontroller is described in Section 3.2, while only the communication with the FPGA is analyzed in this section. The diagram flow on which the C-code of the microcontroller is based is shown in Figure 5.

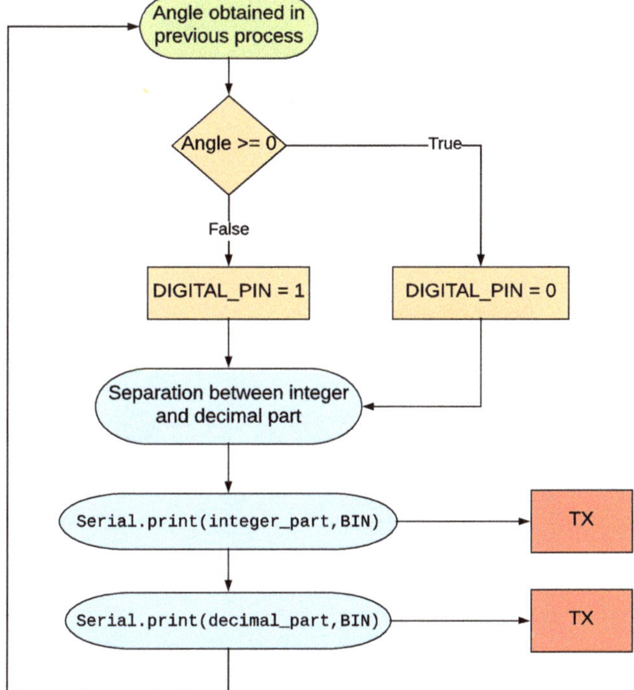

Figure 5. Diagram flow to send angle.

For correct and easy understanding by the FPGA, it is necessary to send the represented angle as several bytes in binary format, not as ASCII codes. The "Serial.println'" Arduino function was discarded, because it used ASCII codes and would even send the comma character to separate the integral and the decimal part. Instead, the Arduino function "Serial.print(Angle)" was used, which sends a binary number through the serial port. The representation and sending of the angle reading was separated into two bytes, as shown in Figure 5. The first byte is the integer part (from 0 to 255) and the second byte is the decimal part (from 0 to 100). No comma is sent over the wire. In FPGA, an input pin will continuously enter data from the transmission pin so that it can make a correct reading of the byte. It is necessary to know:

- When a byte transmission starts;
- When a byte transmission ends;
- When a bit can be captured;
- When the necessary bits are saved in a buffer until the byte is complete.

In order to solve the previous problems and features, an intermediate module in IceStudio (Figure 6) was implemented.

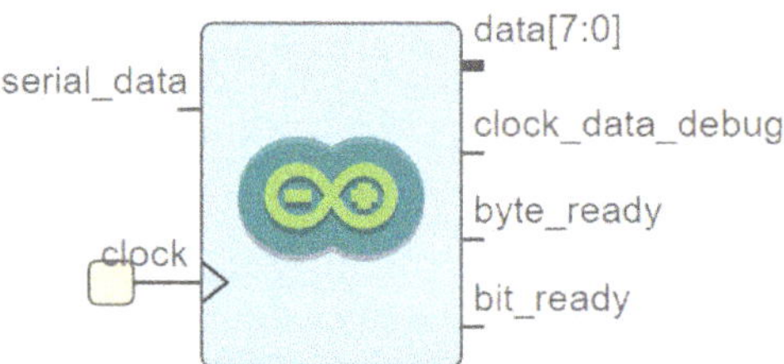

Figure 6. Appearance of Arduino Nano module in IceStudio.

This was implemented in Verilog by two machine states, with their corresponding sensitivity lists and the diagram flow represented in Figure 7.

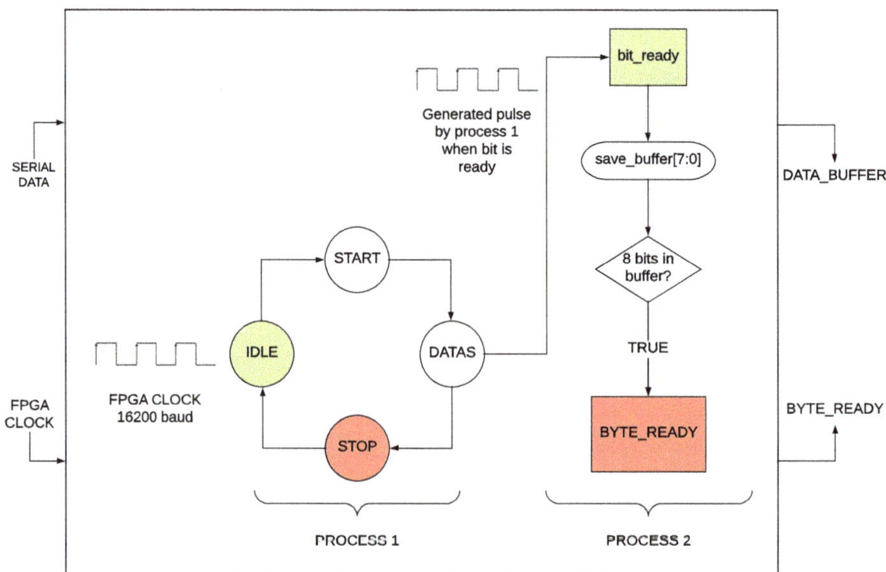

Figure 7. Diagram flow for Arduino interface.

Two distinct processes were used:

Process 1: This process only provides the next system state at the moment at which it can capture a bit and save it in the buffer. Thus, it is necessary to know the speed of the transmission. The states are the following:

- IDLE: The process remains in this state until the transmission starts, which will then lead to the next state (START);
- START: The serial transmission protocol begins with a start condition, and this state will allow recognition of when this condition ends in order to start saving bits in the buffer;
- DATA: Since the transmission speed is already known and the condition of START in the previous state has been recognized, in this state a flag will change its value when the bit is ready to be stored in the buffer, of which Process 2 will be in charge of this storage;
- STOP: In addition to a START condition, the serial transmission protocol used in Arduino has a STOP condition. This state allows recognition of the time Arduino takes to carry out this last condition—it will then return to the first state until a new transaction begins.

Process 2: This process is activated by Process 1. When Process 1 determines that a bit is available on the bus to be captured, it will set a clock flag on, initiating Process 1 through a sensitivity list. An example flow diagram could be:

- Wait until the sensibility list is activated—this will indicate that a bit can be captured;
- Bits will be stored in a buffer forming a byte, which will represent the integer or decimal part of the angle at that moment;
- When the byte is prepared to be captured by two consecutive modules, a channel will be on. Both the outputs and the buffer with the 8 bits and the "byte_ready" channel will be available.

At this point, the FPGA is able to differentiate between when it can capture a byte (BYTE_READY) and from where it has to capture the data bus (DATA_BUFFER). However, an aspect that is not part of the communication itself is that it is important to analyze whether a correct operation is required—that is, if the microcontroller has been previously told to continuously send the integer and decimal part of the angle. If this data is not correctly interpreted, it is possible that an angle on the FPGA is formed by a decimal part of an angle n, and the integer part of the angle n + 1. To do this, a module is created in IceStudio that is capable of ordering these values. The appearance of this module in IceStudio is shown in Figure 8.

Figure 8. Module to arrange data from Arduino.

The final communication system between Arduino and IceZum Alhambra from a POV of the FPGA is represented in Figure 9.

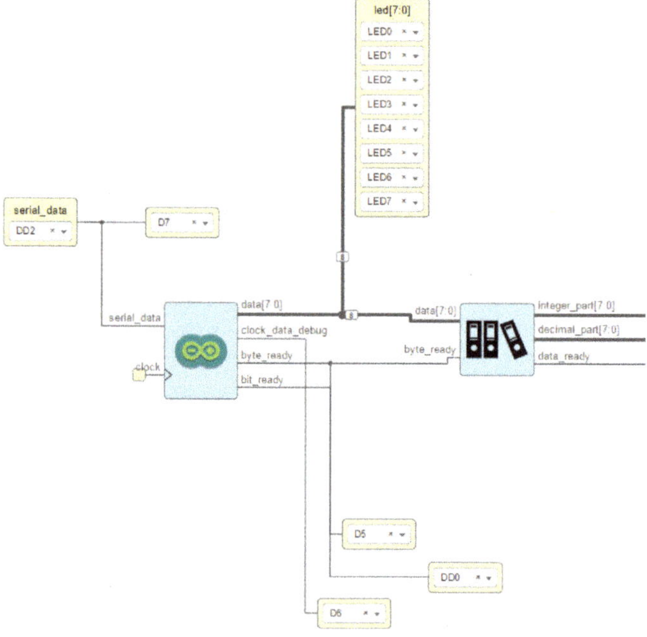

Figure 9. Communication between Arduino and IceZum Alhambra.

3.4. PD Control in FPGA

A PID controller can simply be used to control the stability of the system. One of the facilities provided by this type of controller is the ease of implementation. The flow diagram of the P controller's behavior is shown in Figure 10.

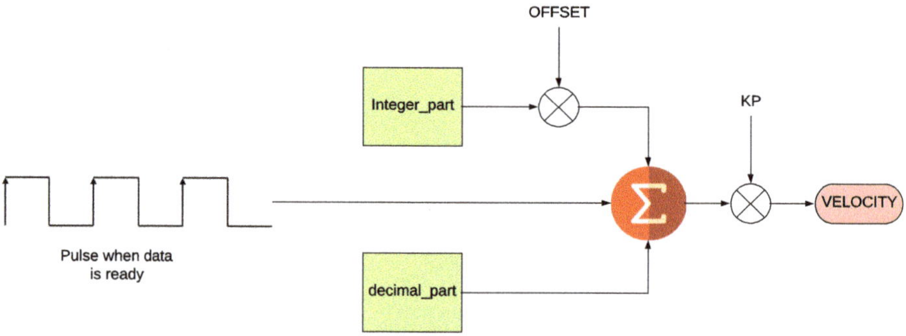

Figure 10. Flow diagram of P control.

The most important features are briefly explained as follows:

- Both the integer part and the decimal part are represented as 8-bit data without a sign. In order to give greater importance to the integer part, there is the option of dividing the decimal part by

100 (Figure 10) or of multiplying the integer part by 100. The first option does not provide good behavior due to the digital treatment of the floating comma. Thus, the second option is preferable.

- The two integer and decimal components are added, and then it is multiplied by a K_p constant, defined as a parameter which can be dynamically changed.

Referring to the D controller, the flow diagram implemented in Verilog is shown in Figure 11.

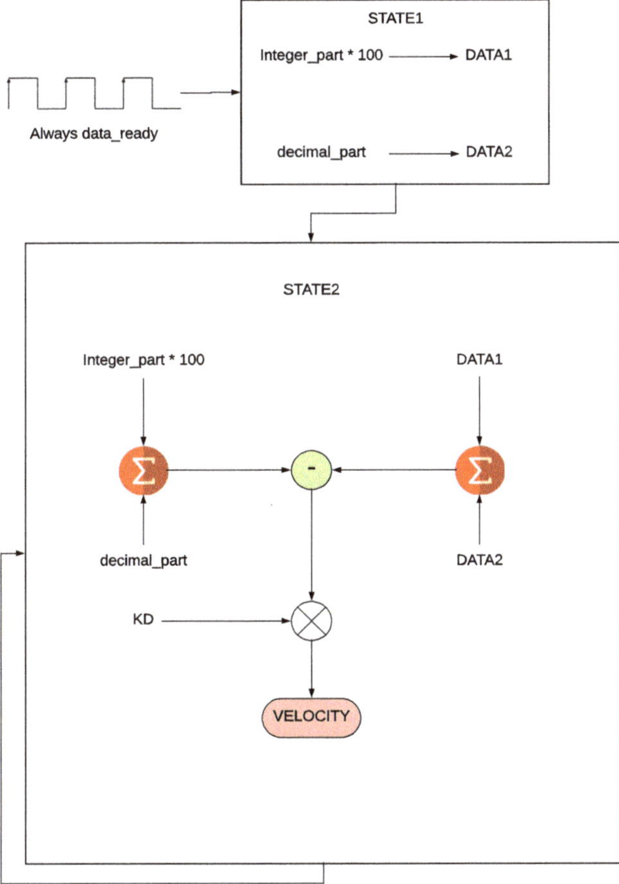

Figure 11. Flow diagram of D control.

Its implementation is composed of a state machine with two states, which will change at each pulse on the *data_ready*. This means that it will change whenever a new angle is available. The D controller is based on its operation on the prediction of future errors. The derivative control action generates a control signal proportional to the derivative of the error signal. A subtraction (derived from the error in time) is therefore carried out between the current error and the last error. Its result is multiplied by the constant K_d. Referring to the closed-loop feedback system, Figure 12 shows the final appearance of the present work developed using IceStudio.

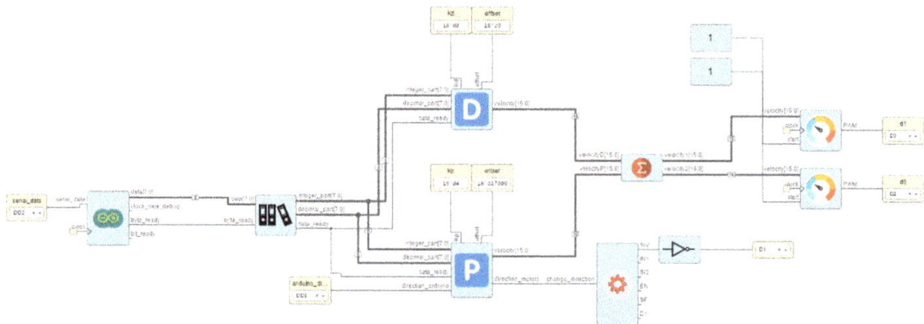

Figure 12. Final appearance of the self-balancing in IceStudio.

3.5. Motor Block in FPGA

In order to correct the current angle and obtain the stabilization, two DC motors were used. The speed of the motors was controlled by a PWM connected to the driver motor through the FPGA. Therefore, a PWM module generator, whose appearance is shown in Figure 13, is needed.

Figure 13. Appearance of PWM module in IceStudio.

Figure 14 shows the block diagram representing its behavior.

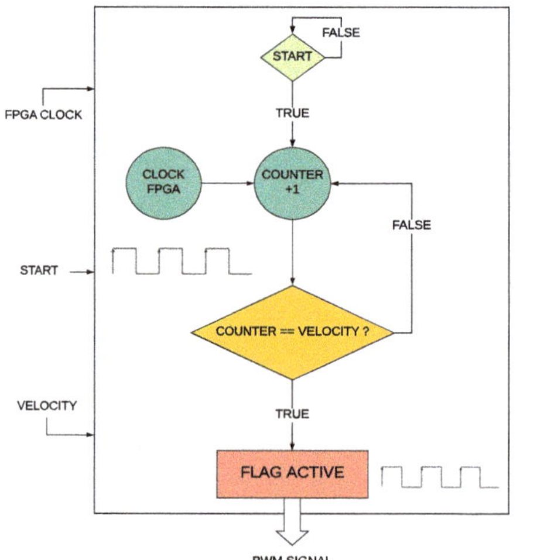

Figure 14. Flow diagram of PWM generator in Verilog.

4. Experiments

In this section, the self-balancing robot will be addressed purely from a hardware perspective describing the chosen physical model and all its components.

4.1. Physical Robot

4.1.1. IceZum Alhambra Board

The Alhambra board (Figure 1) was used as the main board and open FPGA (Section 2.2) to implement all the necessary systems that can be parallelized. For this purpose, the PD control, the calculation of the speed, and the motor control were implemented on this FPGA.

4.1.2. Arduino Nano-Processor

In order to allow a simple implementation of i2c communication with MPU6050 (implementation with FPGA was tested and is described in Section 4.4) and to avoid complex calculations in the FPGA, an ATMEGA microcontroller (Figure 15a) was used. Arduino Nano was chosen to develop the above features.

4.1.3. MPU6050

The MPU6050 (Figure 15b) is an Inertial Measure Unit (IMU) with six degrees of freedom (6DOF) manufactured by Invensense. It has an accelerometer and gyroscope, and allows communication by both SPI and i2c bus. To correct some of the data collection problems, it incorporates an internal processor (Digital Motion Processor, DMP) that executes data fusion algorithms (Motion Fusion) to combine the measurements of the internal sensors, avoiding having to perform the filters externally.

4.1.4. Motor Driver

For the DC motor control, which allows for robot stabilization, the MC33926 was used. It allows control of the speed and direction from up to two motors using a PWM signal which is generated by the module described in Section 3.5. The Figure 15c represents the motor driver.

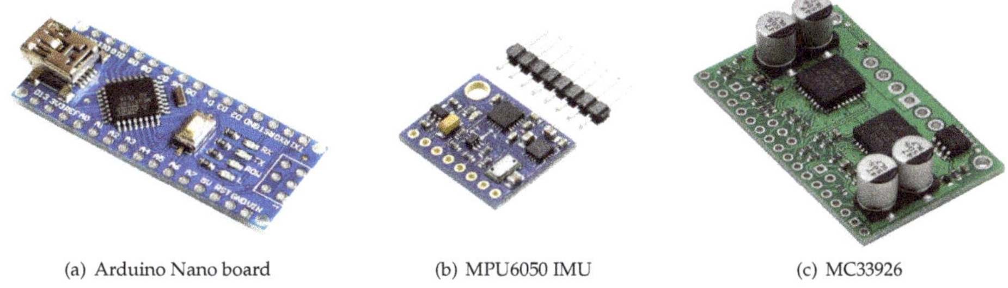

(a) Arduino Nano board   (b) MPU6050 IMU   (c) MC33926

Figure 15. Physical components of the Self-Balancing Robot.

4.1.5. PCB Shield

After implementing the entire system and considering the necessary connection diagram between the microcontroller and FPGA and the motor driver, a printed circuit is advisable to solve some noise problems, the excessive number of cables, etc. A printed circuit board was developed using Altium Designer [38] as the design tool.

Figure 16 shows a 3D representation of the final system with all its added components.

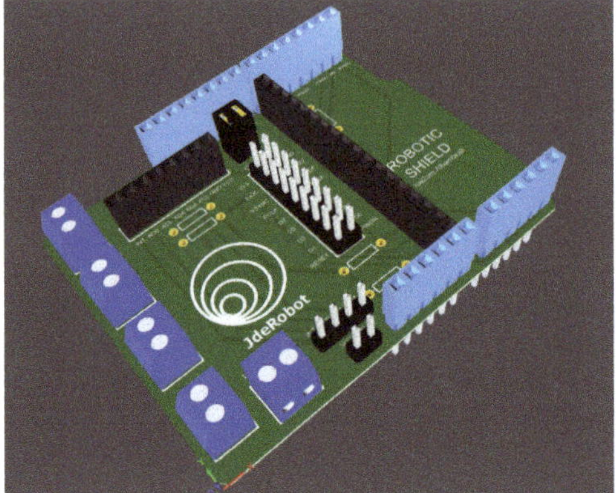

Figure 16. 3D representation of shield for IceZum Alhambra II.

4.2. Inverted Pendulum

Knowing the physics of a self-balancing robot [39] and aiming to solve the classic problem of the inverted pendulum, the mechanical structure of the Figure 17, designed with SolidWorks [39], is proposed to integrate and assemble the rest of the components.

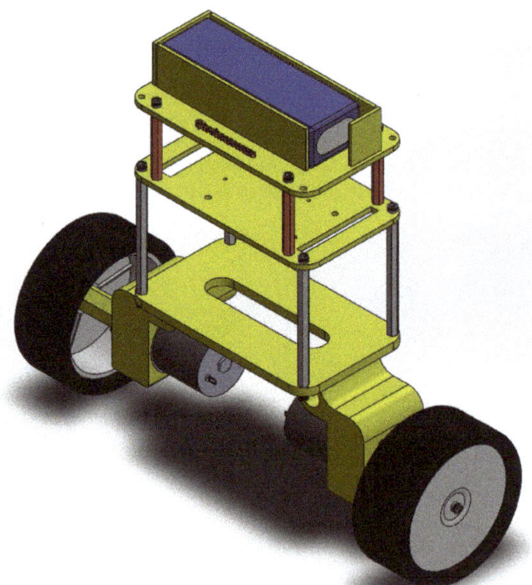

Figure 17. Balancing Robot perspective.

Different aspects of the design of this structure are considered, which are directly related to the physics of a self-balancing robot, and with it, of the inverted pendulum. As mentioned in Section 2.3, a system at rest is stable when its center of mass is closer to the horizontal plane. If we consider that the nature of the proposed system is inherently unstable, it is necessary to know the best point to situate the center of mass in order to provide better stability. According to the mathematical modeling characterization, it is assumed that, in order to achieve greater ease in stabilization, the center of mass should be placed above the midpoint of the vertical axis of our system. Therefore, in order to achieve this positioning, we must consider the weight of all components. In Figure 18, a SolidWorks calculation is represented from this center of mass where only the heavier components of the final system are considered, including DC motors, batteries, mechanical structures, and wheels.

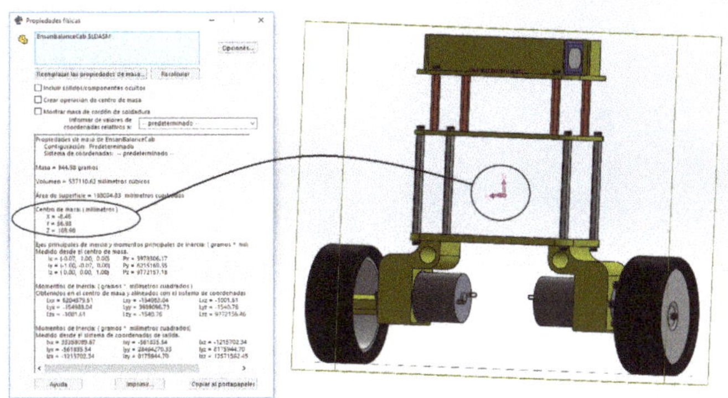

Figure 18. Final system of center of mass.

4.3. Final Results

A set of videos demonstrating the correct behavior in the Self-Balancing robot can be seen at (https://www.youtube.com/watch?v=u-KACjWmcKw). Also, the process through to the end can be found in (https://youtu.be/dQg8NQP7CfQ, https://youtu.be/d_1bnjbpQks, https://youtu.be/mLyxewOVGug). In order to manufacture the mechanical structure, a 3D printer was used (https://youtu.be/rKoIdgaJU2k). The final system is shown in Figures 19 and 20.

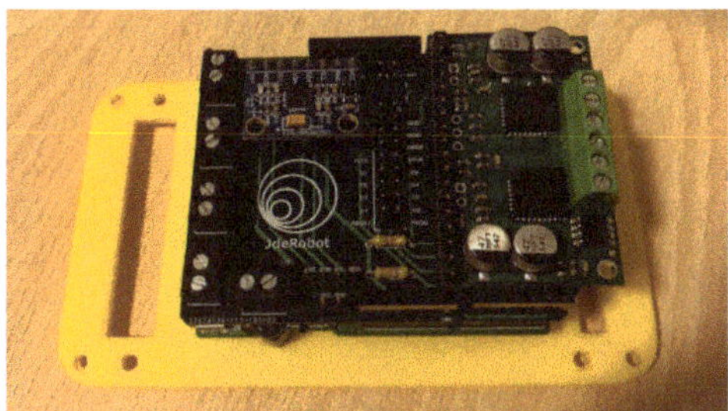

Figure 19. Final system with physical components assembled.

Figure 20. Final results of Self-Balancing Robot.

4.4. Alternative Design without Arduino

The proposed design before adding the microcontroller is shown in Figure 21.

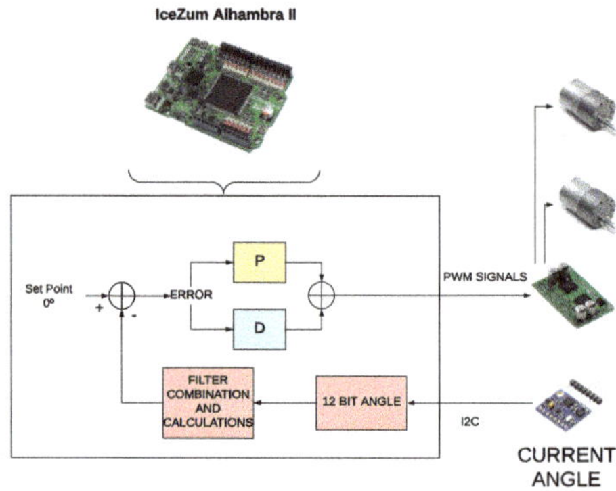

Figure 21. Hardware design of the inverted pendulum control without microcontroller.

The capture of the angle value was implemented in the FPGA. To do this, an i2c module was developed—and this presented a challenge, particularly considering the fact that a state machine and tri-state module were needed. The MPU6050 outputs were reminded without the use of DMP—2 bytes which corresponded with the accelerometer and gyroscope (12 bits for each one). The bytes had to be filtered and carefully treated in order to solve the drift problem and noise. Moreover, these values must be combined to allow reliability in terms of time and to blend the advantages of both sensors. The above development is not feasible with the number of logic gates or with the need to use sine and tangent functions. For this reason, a microcontroller with the i2c incorporated and the possibility of using DMP was clearly the best option.

5. Conclusions

FPGAs are a good intermediate option between microprocessors and ASIC for computing in many technological fields, as they combine the flexibility of software with the high-speed operation of hardware, and can keep costs low. However, most of the FPGA tools are currently proprietary and expensive. The open-source community has developed good FPGA editing and synthesis tools like IceStorm, IceStudio, and the IceZum Alhambra board. Currently supported FPGAs are not yet the most advanced models, but they already allow for the development of interesting robotic applications. A proof-of-concept robotic application was described in the present study—i.e., the inverted pendulum robot. It was fully developed using open FPGA technologies. It includes a perception module, a control module, and a motor module. Perception is based on an inertial IMU sensor. It was first developed with the sensor directly connected to the FPGA board, but there was significant noise in the data from this sensor. Finally, an intermediate Arduino processor was selected to filter out the noise and to send the filtered IMU data to the FPGA board through an i2c connection.

The control module performs a Proportional Derivative (PD) feedback algorithm inside the FPGA board. It feeds the motor drivers with the proper commands to keep the inverted pendulum robot raised and standing up even in the presence of disturbances. The FPGAs allow a new hardware approach to robot programming. Instead of a sequence of instructions, the robot logic is designed naturally in a parallel way by default. The main decomposition of robot tasks is now spatial in the FPGA circuit, which is more than sequential in the processor time. All the modules inside the FPGA hardware run concurrently at a pace of clock frequency. This can be of great use, for instance, in reactive robot behaviors. The hardware allows for continuous control instead of iteration-based software. Regarding future research, the authors are working on programming a drone with a camera to visually follow colored objects in 3D, fully using open FPGA tools. This includes the support for image acquisition directly from the FPGA circuit and the communication with common drone flight controllers (like PX4 or ArduPilot) through PPM encoding. A second consideration to extend the current work is the development of a library of FPGA blocks which can be reused in further robotics applications.

Author Contributions: Conceptualization, J.O., E.C. and J.C.; methodology, E.C. and J.C.; software, J.O.; validation, J.O.; formal analysis, J.O., E.C. and J.C.; investigation, J.O., E.C. and J.C.; resources, J.O. and E.C.; data curation, J.O.; writing–original draft preparation, J.O., E.C. and J.C.; writing–review and editing, E.C. and J.C.; visualization, J.O., E.C. and J.C.; supervision, E.C. and J.C.; project administration, E.C.; funding acquisition, J.C.

Funding: This work was partially funded by the Community of Madrid through the RoboCity2030-III project (S2013/MIT-2748) and by the Spanish Ministry of Economy and Competitiveness through the RETOGAR project (TIN2016-76515-R).

Acknowledgments: The authors wish to thank Prof. Diego P. Morales from the Biochemistry and Electronics as Sensing Technologies Group of the University of Granada for his insightful suggestions and comments on this work.

Conflicts of Interest: The authors declare no conflict of interest.

References

1.   Nickolls, J.; Dally, W.J. The GPU Computing Era. *IEEE Micro.* **2010**, *30*, 56–69, doi:10.1109/MM.2010.41. [CrossRef]

2.   Alkhafaji, F.S.; Hasan, W.Z.; Isa, M.; Sulaiman, N. Robotic Controller: ASIC versus FPGA—A Review. *J. Comput. Theor. Nanosci.* **2018**, *15*, 1–25. [CrossRef]

3.   Sharma, A.K. *Programmable Logic Handbook: PLDs, CPLDs and FPGAs*; McGraw-Hill Handbooks: New York, NY, USA, 1998.

4.   Brown, S.D.; Francis, R.J.; Rose, J.; Vranesic, Z.G. *Field-Programmable Gate Arrays*; The Springer International Series in Engineering and Computer Science; Springer: Berlin, Germany, 1992.

5.   Semiconductor, L. FPGA Lattice. Available online: https://www.latticesemi.com/ (accessed on 10 September 2018).

6.   Xilinx. 2018. Available online: https://www.xilinx.com/ (accessed on 20 September 2018).

7.   Intel. 2018. Available online: https://www.intel.es/content/www/es/es/fpga/devices.html (accessed on 21 December 2018).

8.   Intel. Stratix 10 GX/SX Device Overview. Available online: https://www.altera.com/content/dam/altera-www/global/en_US/pdfs/literature/hb/stratix-10/s10-overview.pdf (accessed on 20 September 2018).

9.   Xilinx. Zynq-7000 All Programmable SoC Data Sheet: Overview. Available online: https://www.xilinx.com/support/documentation/data_sheets/ds190-Zynq-7000-Overview.pdf (accessed on 23 October 2017).

10.   Altera. Nios II Gen2 Processor Reference Guide. Available online: https://www.altera.com/content/dam/altera-www/global/en_US/pdfs/literature/hb/nios2/n2cpu-nii5v1gen2.pdf (accessed on 20 June 2018).

11.   Xilinx. Using the MicroBlaze Processor to Accelerate Cost-Sensitive Embedded System Development. Available online: https://www.xilinx.com/support/documentation/white_papers/wp469-microblaze-for-cost-sensitive-apps.pdf (accessed on 4 September 2018).

12.   RISC-V. Available online: https://riscv.org/ (accessed on 25 December 2018).

13.   Mi-V RISC-V Ecosystem. Available online: https://www.microsemi.com/product-directory/fpga-soc/5210-mi-v-embedded-ecosystem (accessed on 13 November 2018).

14.   Dennis, D.K.; Priyam, A.; Virk, S.S.; Agrawal, S.; Sharma, T.; Mondal, A.; Ray, K.C. Single cycle RISC-V micro architecture processor and its FPGA prototype. In Proceedings of the 2017 7th International Symposium on Embedded Computing and System Design (ISED), Durgapur, India, 18–20 December 2017; pp. 1–5.

15.   Freund, K. "Microsoft: FPGA Wins Versus Google TPUs For AI". Available online: https://www.forbes.com/sites/moorinsights/2017/08/28/microsoft-fpga-wins-versus-google-tpus-for-ai/ (accessed on 24 March 2018).

16.   Ghosh, S. *Hardware Description Languages: Concepts and Principles*; IEEE Computer Society Press: New York, NY, USA, 2000.

17. Nane, R.; Sima, V.M.; Pilato, C.; Choi, J.; Fort, B.; Canis, A.; Chen, Y.T.; Hsiao, H.; Brown, S.; Ferrandi, F.; et al. A survey and evaluation of FPGA high-level synthesis tools. *IEEE Trans. Comput.-Aided Des. Integr. Circuits Syst.* **2016**, *35*, 1591–1604. [CrossRef]

18. Chu, P.P. *RTL Hardware Design Using VHDL: Coding for Efficiency, Portability, and Scalability*; John Wiley & Sons: Hoboken, NJ, USA, 2006.

19. Donald Thomas, P.M. *The Verilog® Hardware Description Language*; Springer Science & Business Media: Berlin, Germany, 2008.

20. SpinalHDL User Guide. Available online: https://spinalhdl.github.io/SpinalDoc/ (accessed on 10 November 2017).

21. Raj, M.D.; Gogul, I.; Thangaraja, M.; Kumar, V.S. Static gesture recognition based precise positioning of 5-DOF robotic arm using FPGA. In Proceedings of the 2017 Trends in Industrial Measurement and Automation (TIMA), Chennai, India, 6–8 January 2017; pp. 1–6. [CrossRef]

22. Zhang, Z.; Xin, Y.; Liu, B.; Li, W.X.Y.; Lee, K.H.; Ng, C.F.; Stoyanov, D.; Cheung, R.C.C.; Kwok, K.W. FPGA-Based High-Performance Collision Detection: An Enabling Technique for Image-Guided Robotic Surgery. *Front. Robot. AI* **2016**, *3*, 51. [CrossRef]

23. Vachhani, L.; Mahindrakar, A.D.; Sridharan, K. Mobile Robot Navigation Through a Hardware-Efficient Implementation for Control-Law-Based Construction of Generalized Voronoi Diagram. *IEEE/ASME Trans. Mechatron.* **2011**, *16*, 1083–1095. [CrossRef]

24. Eteokleous, N.; Ktoridou, D. Educational robotics as learning tools within the teaching and learning practice. In Proceedings of the 2014 IEEE Global Engineering Education Conference (EDUCON), Istanbul, Turkey, 3–5 April 2014; pp. 1055–1058. [CrossRef]

25. Khatib, B.S.O. *Springer Handbook of Robotics*; Springer: Berlin, Germany, 2016.

26. Kung, Y.S.; Shu, G.S. Development of a FPGA-based motion control IC for robot arm. In Proceedings of the 2005 IEEE International Conference on Industrial Technology, Hong Kong, China, 14–17 December 2005; pp. 1397–1402. [CrossRef]

27. Nema, R.; Thakur, R.; Gupta, R. Design & Implementation of PID Controller Based On FPGA with PWM Modulator. *Int. J. Soft Comput. Eng. IJSCE* **2013**, *3*, 2231–2307.

28. Linares, J.C.; Barrientos, A.; Márquez, E.M. Hybrid Bio-Inspired Architecture for Walking Robots Through Central Pattern Generators Using Open Source FPGAs. In Proceedings of the 2018 IEEE/RSJ International Conference on Intelligent Robots and Systems (IROS), Madrid, Spain, 1–5 October 2018; pp. 7071–7076. [CrossRef]

29. Arduino. Available online: https://www.arduino.cc/ (accessed on 26 May 2017).

30. Romanov, A.; Bogdan, S. Open source tools for model-based FPGA design. In Proceedings of the 2015 International Siberian Conference on Control and Communications (SIBCON), Omsk, Russia, 21–23 May 2015; pp. 1–6.

31. Wolf, C.; Lasser, M. Project Icestorm. Available online: http://www.clifford.at/icestorm (accessed on 15 January 2018).

32. Tarjeta IceZum Alhambra II. Available online: https://alhambrabits.com/alhambra/ (accessed on 20 January 2018).

33. IceStudio. Available online: https://icestudio.readthedocs.io/en/latest/ (accessed on 20 January 2018).

34. Romanov, A.; Romanov, M.; Kharchenko, A. FPGA-based control system reconfiguration using open source software. In Proceedings of the 2017 IEEE Conference of Russian Young Researchers in Electrical and Electronic Engineering (EIConRus), St. Petersburg, Russia, 1–3 February 2017; pp. 976–981. doi:10.1109/EIConRus.2017.7910719. [CrossRef]

35. Pathak, K.; Franch, J.; Agrawal, S.K. Velocity and position control of a wheeled inverted pendulum by partial feedback linearization. *IEEE Trans. Robot.* **2005**, *21*, 505–513. [CrossRef]

36. Orozco, L.M.L.; Lomeli, G.R.; Moreno, J.G.R.; Perea, M.T. Identification Inverted Pendulum System using Multilayer and Polynomial Neural Networks. *IEEE Latin Am. Trans.* **2015**, *13*, 1569–1576. [CrossRef]

37. Yu, L.H.; Jian, F. An Inverted Pendulum Fuzzy Controller Design and Simulation. In Proceedings of the 2014 International Symposium on Computer, Consumer and Control, Taichung, Taiwan, 10–12 June 2014; pp. 557–559. [CrossRef]

38. Altium Designer. Available online: https://www.altium.com/altium-designer/ (accessed on 19 June 2018).
39. SolidWorks. Available online: https://www.solidworks.com/es (accessed on 25 October 2018).

Article

A New Seven-Segment Profile Algorithm for an Open Source Architecture in a Hybrid Electronic Platform

José R. García-Martínez [†], Juvenal Rodríguez-Reséndiz * and Edson E. Cruz-Miguel [†]

División de Investigación y Posgrado, Facultad de Ingeniería, Universidad Autónoma de Querétaro (UAQ), Cerro de las Campanas, S/N, Col. Las Campanas, Querétaro C.P. 76010, Mexico; jose.gm@uaq.mx (J.R.G.-M.); ecruz30@alumnos.uaq.mx (E.E.C.-M.)
* Correspondence: juvenal@uaq.edu.mx; Tel.: +52-442-192-1200
† These authors contributed equally to this work.

Received: 28 April 2019; Accepted: 6 June 2019; Published: 10 June 2019

Abstract: The velocity profiles are used in the design of trajectories in motion control systems. It is necessary to design smoother movements to avoid high stress in the motor. In this paper, the rate of change in acceleration value is used to develop an S-curve velocity profile which presents an acceleration and deceleration stage smoother than the trapezoidal velocity profile reducing the error at the end of the duty-cycle pre-established in one degree of freedom (DoF) application. Furthermore, a new methodology is developed to generate a seven-segment profile that works with negative velocity and displacement constraints applying an open source architecture in a hybrid electronic platform compounded by a system on a chip (SoC) Raspberry Pi 3 and a field programmable gate array (FPGA). The performance of the motion controller is measured through the comparison of the error obtained in real-time application with a trapezoidal velocity profile. As a result, a low-cost platform and an open architecture system are achieved.

Keywords: low-cost platform; FPGA; S-curve; motion control; robotics; SoC

---

1. Introduction

The velocity profiles have been studied broadly in recent years to design point-to-point trajectories in robot manipulators, conveyor belts, computer numerical control (CNC) machinery or whatever system with the use of direct current (DC) and alternating current (AC) motors [1,2]. Velocity profiles have an essential role in motion control since it is possible to accomplish a target position reducing the vibrations and the energy consumption, increasing the precision and the durability of the systems [3,4]. Nowadays, a great variety of velocity profiles exist, but their accuracy depends on the velocity's demeanor. Since, if the velocity changes abruptly, the behavior of acceleration could cause discontinuities in the trajectory [5]. The rate of change in acceleration is denominated as jerk [6], whether acceleration changes too fast, the vibrations increase their frequency causing damage to the structure of the mechanical system [7]. Therefore, when something like this happens, it is possible to deduce that the jerk value is too big, which means that the energy consumption is high.

There are different velocity profiles applied to specific processes, the common ones are the triangular velocity profile, parabolic velocity profile and the trapezoidal velocity profile [8]. The triangular velocity profile is a piece-wise defined function given by two linear segments corresponding to the acceleration and deceleration of the actuator, the jerk value is high because of the acceleration changes radically. On the other hand, the parabolic velocity profile presents a smoother velocity curve than the triangular, making the jerk value less than triangular profile but neither of them maintain a velocity constant phase [9,10], it means that they accelerate and decelerate the actuator immediately. The trapezoidal velocity profile consists of three phases: acceleration, constant velocity and deceleration phase [10,11]. A constant velocity phase offers less wear on the actuator extending its life period, since, the change in acceleration occurs after a period and not abruptly after to reach the desired velocity [12,13]. The trapezoidal velocity profile is the most used in the industry, although the jerk value is high-rise [5,14]. A second-degree polynomial describes the transition of the position making accessible the implementation in an embedded system due to the low level of processing [15,16]. On the other hand, the seven-segment velocity profile, known as the S-curve velocity profile, has been studied broadly in recent years obtaining better results than other profiles because of the jerk value takes a constant amount [17], decreasing the damage produced by high-frequency vibrations in the structure of the system. The implementation in an embedded system can need a complex architecture to support the seven position's equations defined by a third-degree polynomial.

The controller is fundamental to the application of a motion control system [18,19], the velocity profile contributes with several points that describe the path, but the controller must follow, reducing the error significantly, the trajectory [20,21]. According to a comparison presented in [2], the controller and the profile generator algorithm can be chosen by the designer according to its experience. For instance, Jeong et al. [22] proposed an algorithm that can determine the coefficients of jerk limited profiles, but it only works with non-negative velocity and displacement constraints. Wang Bangji et al. [23] developed a velocity profile algorithm for stepper motor controller in an field programmable gate array (FPGA) where the characteristics of the stepper motor are introduced by the user to generate the velocity profile. The algorithm is restricted to other motors. Tou Wai Kei et al. [24] designed a speed regulator by implementing an S-curve velocity profile in a microcontroller to control an elevator featuring direct landing. The profile generator only accepts the maximum acceleration, an initial jerk value, and a maximum velocity. Working with microcontrollers can presents problems at the moment to migrate the algorithm to a different family of microcontrollers due to the internal architecture used among them, in the FPGA the code is preserved, it does not change and it can be migrated in an easy way without altering the algorithm. This paper presents a new methodology to obtain the S-curve coefficients in real-time, which includes the non-negative velocity and displacements constraints applied to DC motors using an open architecture based on a Raspberry Pi 3 combined with an FPGA which is a low-cost platform compared to closed architectures on the market capable to generate trajectories. Furthermore, the user can introduce the total displacement, the duration of the movement and the length of the acceleration-deceleration phases. Besides, the profile generator calculates the acceleration and jerk parameters according to the desired position.

2. Background

2.1. General Model of the Seven-Segment Velocity Profile

The motion planning designer must develop smoother trajectories to avoid discontinuities in the acceleration and reduce the strain and exertion on the actuators and the mechanical architecture [25,26]. Since a third-degree polynomial models the behavior of the position, the seven-segments velocity profile offers the possibility to maintain the jerk with a constant value, obtaining a step profile for the rate of change in acceleration [27]. An important aspect to mention is that, when the degree of the polynomial is increased or decreased, the profiles tend to shift themselves into a particular position, adopting a different shape. For instance, in Figure 1c, the acceleration has taken the shape of a trapezoidal profile allowing linear changes at the acceleration, even the velocity adopts a continuous shape connected by parabolic-blends in the form of S-curve for the acceleration phase and an inverse S-curve for the deceleration stage.

It is necessary to analyze Figure 1 to compute the values of the u_0, u_1, u_2 and u_3 parameters of Equation (1), by seven-segments with an interpolation method [1]; each segment represents an equation depending on the motion profile under analysis. So, it is acceptable to propose a constant value for the jerk according to a given acceleration, but it must exist a relationship with the maximum amount of the speed given by the data-sheet of the actuator.

$$\theta_d(t) = a_0 + a_1 t + a_2 t^2 + a_3 t^3, \tag{1}$$

where θ_d is the aim position. When the maximum values of the jerk, acceleration, and velocity are known, the Equations (3)–(5) can be applied to generate the wished trajectory with the seven-segments velocity profile. The total duration of the movement T is also known, and it is provided by the designer. As mentioned before, the jerk has a step profile because it retains its value as a constant, so that in Figure 1d one can see how it varies with respect to the time and get (2).

$$J(t) = \begin{cases} j_{max} & t \in [0, T_{s1}) \\ 0 & t \in [T_{s1}, T_{s2}) \\ j_{min} & t \in [T_{s2}, T_{s3}) \\ 0 & t \in [T_{s3}, T_{s4}) \\ j_{max} & t \in [T_{s4}, T_{s5}) \\ 0 & t \in [T_{j5}, T_{s6}) \\ j_{min} & t \in [T_{s6}, T]. \end{cases} \tag{2}$$

Assuming that $j_{min} = -j_{max}$ in four periods, T_{si} is the length of the i-th segment, where $i = 0, 1, 2, ..., 6$. The acceleration is calculated after integrating (2), segment by segment. Equation (3) is the general equation to get acceleration. Equation (4) is the equation of the velocity and (5) is the equation to obtain the position.

$$\alpha(t) = \alpha(T_{si}) + \int_{T_{si}}^{T} J(\tau_i) d\tau_i \tag{3}$$

$$\omega(t) = \omega(T_{si}) + \int_{T_{si}}^{T} \alpha(\tau_i) d\tau_i \tag{4}$$

$$\theta(t) = \theta(T_{si}) + \int_{T_{si}}^{T} \omega(\tau_i) d\tau_i. \tag{5}$$

The relative time parameter of the integral is defined as $\tau_i = T - T_{si}$ where $i = 1, 2, 3, ..., 6$ represents the segments of the displacement. The result of the integration of (3) is the acceleration profile showed in Figure 1c. The acceleration shows a linear variation until reach a constant value, and then presents a linear deceleration.

In order to draw the acceleration profile with the desired characteristics, it is necessary to substitute the a_{acc} and the j_{max} values in (6), which is the result of the integration of (3). Notice that the acceleration phase goes from the origin to T_{s3} in Figure 1c, whereas the deceleration phase $(T - T_{s4})$ is compound by an inverse trapezoid.

$$\alpha(t) = \begin{cases} j_{max} T_{s1} & t \in [0, T_{s1}) \\ a_{acc} & t \in [T_{s1}, T_{s2}) \\ a_{acc} + j_{min}(T_{s3} - T_{s2}) & t \in [T_{s2}, T_{s3}) \\ 0 & t \in [T_{s3}, T_{s4}) \\ j_{max}(T_{s5} - T_{s4}) & t \in [T_{s4}, T_{s5}) \\ a_{dec} & t \in [T_{s5}, T_{s6}) \\ a_{dec} + j_{min}(T - T_{s6}) & t \in [T_{s6}, T] \end{cases} \tag{6}$$

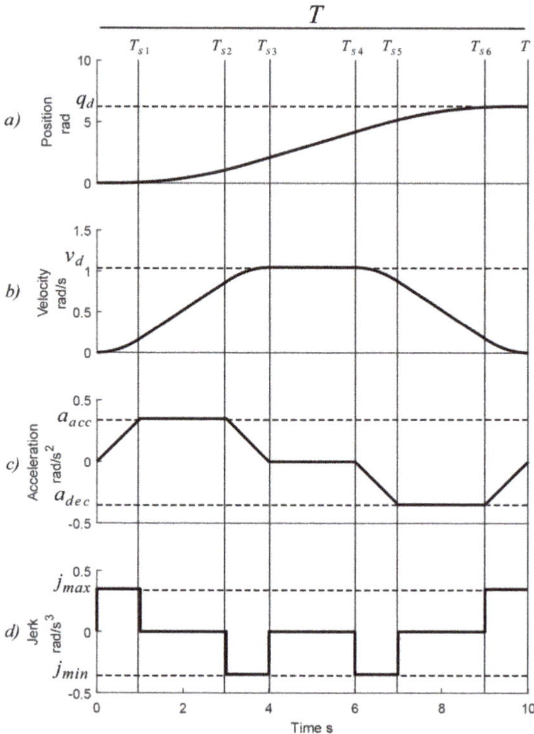

Figure 1. Motion profiles by given a set-point, (**a**) position, (**b**) velocity, (**c**) acceleration and (**d**) jerk.

According to (4), it is possible to compute the velocity profile of each segment from the integral of the acceleration. The seven-segments velocity profile is modeled by (7).

$$\omega(t) = \begin{cases} v_1 + \frac{j_{max}}{2}(\tau_1)^2 & t \in [0, T_{s1}) \\ v_1 + \frac{j_{max}}{2}T_1^2 + a_{acc}\tau_2 & t \in [T_{s1}, T_{s2}) \\ v_1 + \frac{j_{max}}{2}T_1^2 + a_{acc}T_2 + a_{acc}\tau_3 + \frac{j_{min}}{2}\tau_3^2 & t \in [T_{s2}, T_{s3}) \\ v_{max} & t \in [T_{s3}, T_{s4}) \\ v_{max} - \frac{j_{max}}{2}(\tau_5)^2 & t \in [T_{s4}, T_{s5}) \\ v_{max} - \frac{j_{max}}{2}T_5^2 - a_{dec}\tau_6 & t \in [T_{s5}, T_{s6}) \\ v_{max} + \frac{j_{max}}{2}T_5^2 - a_{dec}T_6 + a_{dec}\tau_7 + \frac{j_{min}}{2}(\tau_7)^2 & t \in [T_{s6}, T]. \end{cases} \tag{7}$$

Here, $\tau_{i-1} = T - T_{si}$ is the relative time of each segment of (7), $i = 1, 2, ...7$ defines the segment under analysis and T_i is the duration of the i-th stage. The acceleration, when the maximum speed v_{max} is constant, turns to zero. Finally, it is necessary to solve (5), to compute the position profile for the i-th segment according to Figure 1a. Equation (8) represents the piece-wise behavior of the position respect the time. Notice that it posses a third-degree polynomial equations that modeled the desired position in certain segments.

$$\theta(t) = \begin{cases} q_1 + v_1\tau_1 + \frac{j_{max}}{6}(\tau_1)^3 & t \in [0, T_{s1}) \\ q_2 + v_2\tau_2 + \frac{a_{acc}}{2}(\tau_2)^2 & t \in [T_{s1}, T_{s2}) \\ q_3 + v_3\tau_3 + \frac{a_{acc}}{2}(\tau_3)^2 - \frac{j_{min}}{6}(\tau_3)^3 & t \in [T_{s2}, T_{s3}) \\ q_4 + v_4\tau_4 & t \in [T_{s3}, T_{s4}) \\ q_5 + v_5\tau_5 - \frac{j_{max}}{6}(\tau_5)^3 & t \in [T_{s4}, T_{s5}) \\ q_6 + v_6\tau_6 - \frac{a_{dec}}{2}(\tau_6)^2 & t \in [T_{s5}, T_{s6}) \\ q_7 + v_7\tau_7 + \frac{a_{dec}}{2}(\tau_7)^2 - \frac{j_{min}}{6}(\tau_7)^3 & t \in [T_{s6}, T], \end{cases} \tag{8}$$

where

$$\begin{cases} q_2 = q_1 + v_1 T_1 + \frac{j_{max}}{6} T_1^3 \\ q_3 = q_2 + v_2 T_2 + \frac{a_{acc}}{2} T_2 \\ q_4 = q_3 + v_3 T_3 + \frac{a_{acc}}{2} T_3^2 - \frac{j_{min}}{6} T_3^3 \\ q_5 = q_5 + v_5 T_5 - \frac{j_{max}}{6} T_5^3 \\ q_6 = q_5 + v_5 T_5 - \frac{j_{max}}{6} T_5^3 \\ q_7 = q_6 + v_6 T_6 - \frac{a_{dec}}{2} T_6^2, \end{cases}$$

when the motor has reached the maximum velocity v_{max} using the desired position θ_d and duration of the movement T proposed, it maintains a constant speed and the position in that phase is a third-degree polynomial.

2.2. Proposed Method to Compute the Desired Jerk

It was necessary to get access to the data-sheet of the actuator to design a point-to-point trajectory [1], and check for some specific characteristics as torque, minimum and maximum values for the velocity and acceleration, and so on [28,29]. These parameters can be used by the designer for solving a determined task using delimiter parameters to prevent damage caused under dynamic loads initiated by molecular bond separation in the material, and to reduce the vibrations on the actuator [25,30,31]. The aim of working with a range of velocities was to assign the desired speed and compute the needed values of the acceleration and the jerk respect to the rate of change in position parameter proposed. A condition to satisfy is presented as follow.

$$\omega_d = \omega(T_{si}) + \int_{T_{si}}^{T} \alpha(\tau_i)d\tau_i \leq \omega_{max}, \tag{9}$$

where ω_d and ω_{max} are scalar of the desired and the maximum speed permitted by the DC motor respectively and they must satisfy (9). Whether a more significant value is declared for ω_d than ω_{max}, the actuator is going to try to reach that speed, demanding a higher voltage than the provided by the manufacturer. Therefore, an approximation for the planning of the seven-segments velocity profile consists of defining the values for the rate of change of position and acceleration over a function based on the acceleration-deceleration stage proposed. The total time for acceleration phase is represented in (10).

$$T_{acc} = T_{s1} + T_{s2} + T_{s3} \tag{10}$$

The phase when the speed is varying respect the time is constituted by three segments, two parabolic-blends and a linear displacement related to the speed profile. Assuming that the acceleration phase is symmetric respect the deceleration phase, it is possible to suppose that $T_{acc} = T_{dec}$. In order to compute T_{acc} it is necessary to multiply the total time of motion T by a factor $\gamma \in R$, where $0 \leq \gamma \leq \frac{1}{2}$. So, the acceleration time can be calculated using $T_a = \gamma T$. A value of acceleration and jerk have to be calculated to use Equations (6)–(8),. The velocity obtained by the desired position can be computed using (11).

$$\omega_d = \omega(T_{s3}) + \int_{T_{s3}}^{T_{s4}} \alpha(\tau_3) d\tau_3 = \frac{\theta_d}{(1-\gamma)T}. \tag{11}$$

The result of (12) took the value of the ω_d as the maximum velocity computed with the desired position θ_d. Therefore, $\omega_d = \omega_{max}$ in the segment $\{T_{s3}, T_{s4}\}$ over Figure 1b. The parameter θ_d is an scalar and its range is (R^-, R^+). The segment of time when the jerk kept a constant amount T_{jerk} is less than the acceleration phase T_{acc}, it means $T_{acc} \geq T_{jerk}$. It must exist a relationship between the T_{acc} and T_{jerk} to ensure the continuity of Equations (6)–(8), so that, the length of the acceleration phase has to be multiplied by a factor $\varphi \in R$. φ can take values in the range $0 \leq \varphi \leq \frac{1}{2}$, thus, the constant value of the jerk should endure $T_{jerk} = \varphi T_{acc}$. Supposing that the acceleration phase is symmetric to the deceleration phase, they have the same duration $T_{acc} = T_{dec}$ with an opposite magnitude, solving (12) to determine the acceleration value.

$$\alpha_d = \alpha(T_{s1}) + \int_{T_{s1}}^{T_{s2}} J(\tau_1) d\tau_1 = \alpha(T_{s5}) + \int_{T_{s5}}^{T_{s6}} J(\tau_5) d\tau_5 = \frac{\theta_d}{\gamma(1-\gamma)(1-\varphi)T^2}. \tag{12}$$

where α_d is the maximum value reached with the desired position parameter in total duration motion proposed. The α_d is a constant parameter, $\alpha_d \in R$. The jerk value can be computed by (13).

$$J_d = \frac{\alpha_d}{T_{jerk}} = \frac{\theta_d}{T_{jerk}\gamma(1-\gamma)(1-\varphi)T^2}. \tag{13}$$

Here J_d is the constant jerk value, $J_d \in R$. Notice that exists a dependence among the values of the velocity ω_d, acceleration α_d and jerk J_d, respect to the total time of the motion an the target position. Once the jerk is obtained, it can be substituted in (2) to draw the jerk profile. The desired position, using (8), can be computed with Equation (14).

$$\theta_d = q_6 + v_6 T_6 + v_7 T - \frac{a_{dec}}{2} T_6^2 + \frac{a_{dec}}{2}(T)^2 - \frac{j_{min}}{6}(T)^3. \tag{14}$$

The value of θ_d is calculated from the last stage of (8) and represents the total displacement from the movement. It is necessary to calculate a negative acceleration and jerk parameters to compute the negative velocity constraints, using Equations (13) and (14). $J_{inv} = sgn(j_{si} - J_d)\,|\,J_d\,|$ indicates if the sign of the jerk profile is positive or negative. Whether the jerk value is negative, the jerk profile presented in Figure 1d changes its shape to Figure 2b. On the other hand, inverting the jerk profile, one can obtain a negative acceleration since the jerk is negative. So, the stage sign for the acceleration can be obtained using the following constraint $\alpha_{inv} = sgn(j_{si} - J_d)\,|\,\alpha_d\,|$, where a negative sign for the acceleration profile correspond to invert the trapezoidal shape of acceleration, see Figure 2a. Using the negative acceleration, the velocity profile presents an inverted S-curve, Figure 3b.

Equations (2)–(8) are inverted after computing the desired jerk and the maximum acceleration needed to reach the jerk condition. It is added a new variable denominated as the initial position θ_0 to compensate the total displacement writing in (15). Once the jerk and acceleration are computed for a negative displacement, in accordance with the actual value, it is possible to use the negative value of the computed jerk to generate the inverse trajectory.

$$\theta_{d'} = \theta_0 + \theta_d, \tag{15}$$

where $\theta_{d'}$ is the final displacement if the initial position θ_0 is the last movement reached for the shaft of the motor, it means that the new desired position has moved in an opposite way, and the total displacement computed from the home point is described in (16):

$$\theta_d = q_0 - q_6 - v_6 T_6 - v_7 T + \frac{a_{dec}}{2} T_6^2 - \frac{a_{dec}}{2}(T)^2 + \frac{j_{min}}{6}(T)^3 \tag{16}$$

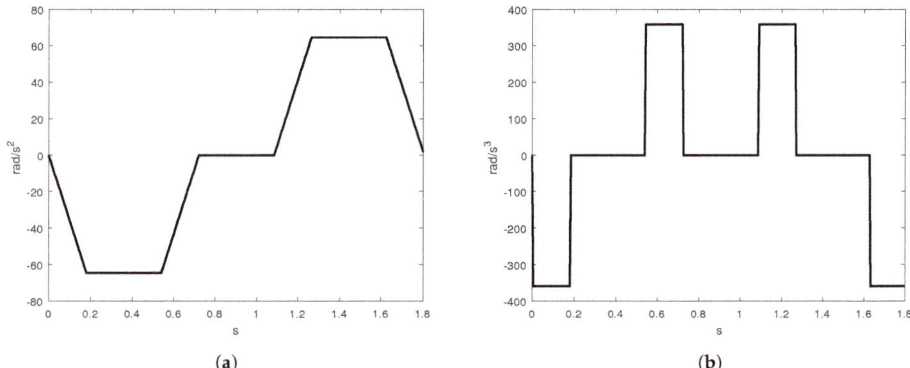

Figure 2. (**a**) Inverse trapezoidal acceleration profile, (**b**) inverse jerk profile.

The negative position is used to return the shaft of the motor to the initial position or to move it in an opposite way to the positive axis, the position profile is presented in Figure 3a.

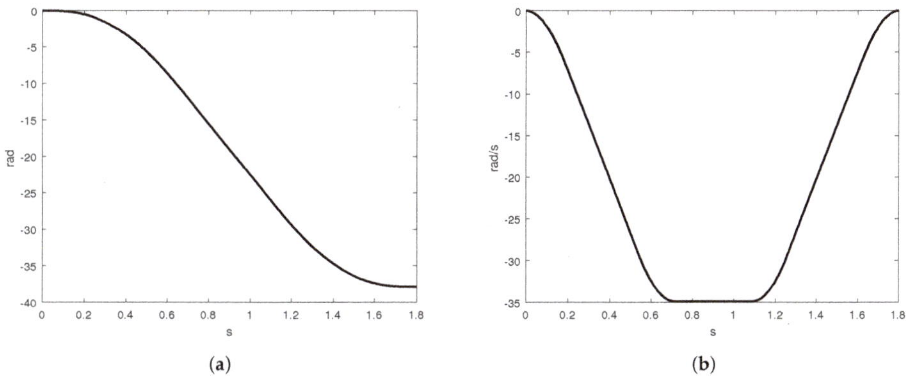

Figure 3. Inverse S-curve velocity profile (**a**) position, (**b**) S-curve velocity profile.

3. Methods and Experimentation

Motion control applications encompass an extensive range of topics related to the control of electromechanical systems, trajectories design to reduce the error increasing the precision of the system and handle of vibrations to minimize damage in mechanical structures. A new methodology is proposed in Figure 4 to reduce the error obtained experimentally applying a motion profile. In order to design a motion controller using a hybrid system compounded by the interaction of a Raspberry Pi 3 and an FPGA ZYBO-ZYNQ XC7Z010 from the family XILINX.

The idea of designing a motion controller with those architectures arises from the technological advances in industry. Using a single-board computer reduces the work-space and increases the possibility about everybody can interact with the system, making interactive the process to the operator in charge of the mechanical system. The processing and control systems are embedded on the Raspberry Pi 3. For the power stage, a pulse-width-modulation (PWM) servo drive model 12A8 from the family advanced motion Control is used [32].

The seven-segment velocity profile is developed in C programming; the user can set the target position θ_d, the total time of displacement T, and the length of the acceleration-deceleration phases to compute the speed needed to reach the set-point and the jerk value.

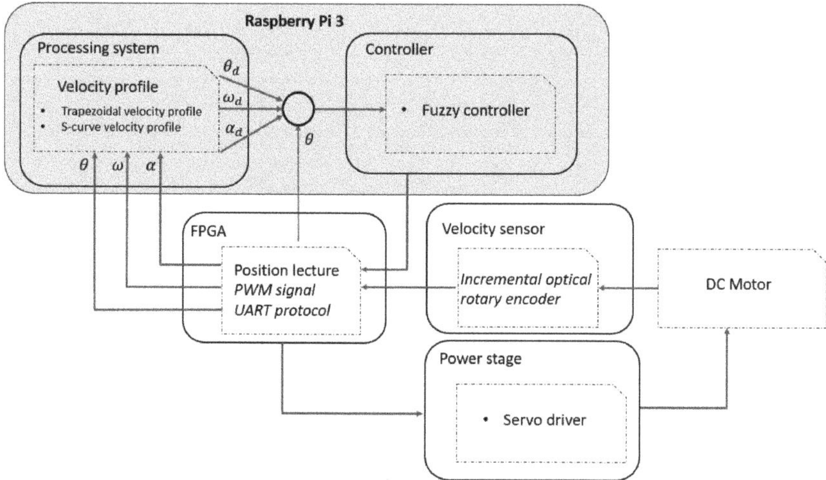

Figure 4. Hybrid electronic topology based on field programmable gate array (FPGA)-system on a chip (SoC).

On the other hand, the FPGA reads the real position of the shaft using a rotary encoder (2000 pulses per revolution). A quadrature signal was monitored each 20 na and it was stored in a register each 5 ms to send the data directly to the interface using the universal asynchronous receiver-transmitter (UART) protocol, Figure 5 shows the sequential logic implemented in the FPGA. $DATA\_Tx$ is the data to send, compounded by 8 bits; for this case, the encoder counts have a bandwidth of 16 bits. Rx and Tx are the communication lines, BaudRate (bits per second) is the transmission speed, it is important configure the same speed at 115,200 bauds in the C program of the Raspberry Pi 3. $DATA\_Rx$ is a buffer where the control signal is received, eo_Tx and eo_Rx are flags that indicate the end of transmission and reception of 1 byte, respectively.

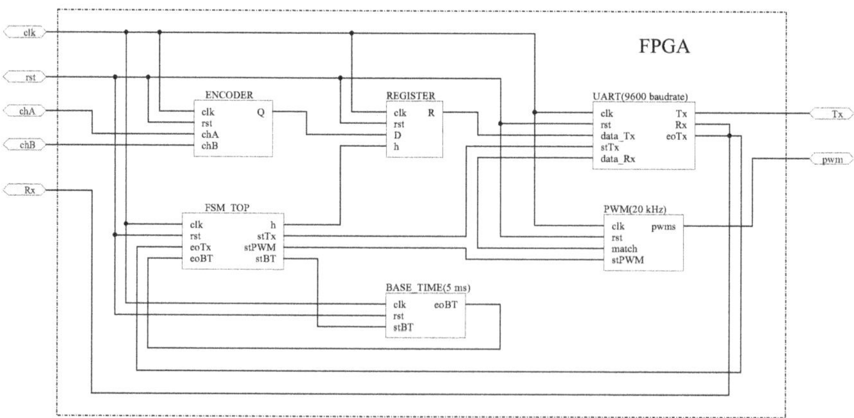

Figure 5. Entities embedded on the FPGA.

The Raspberry Pi 3 generated the trajectory parameters in real-time, using a sample time $T_s = 5$ ms, and sent them to the controller to minimize the error. The control signal was transmitted through the general purpose input–ouput (GPIO's) of the Raspberry Pi 3 to the FPGA in the form eight bits of information, the FPGA receives the data and transforms the control signal into PWM signal in order to send it to the servo drive to control the position of the motor.

S-Curve Velocity Profile Parameters

The S-curve velocity profile implementation was developed using (2)–(5). There were several considerations at the moment to define a trajectory such as: (a) total duration for the acceleration-deceleration must be equal for both stages $T_{acc} = T_{dec}$, (b) the magnitude of the acceleration α_d was obtained by the desired position θ_d, (c) the jerk value is obtained with the α_d magnitude and θ_d, all those values are calculated with the total duration of the movement T.

The parameters proposed for the design of the S-curve implementation are $\theta_d = 12\pi$ *rad* and $T = 1.8$ s, notice that the speed is calculated from the jerk value or using (9). Once the length of the movement was known, the acceleration-deceleration time was computed. For this application, the total time of the acceleration is divided by $\gamma = \frac{4}{10}$, and $T_{acc} = \gamma T$ in order to have a symmetric profile. On the other hand, the duration of the jerk stage must be divided by four times the acceleration phase T_{acc} to ensure an S-curve velocity profile equal in length of acceleration, maximum velocity and deceleration stages, the proportional value was $\varphi = \frac{1}{4}$, so that the jerk was going to remain zero for $T_{jerk} = \frac{2}{4}T_{acc}$. The constant velocity stage had a duration of $\frac{2}{10}T$. The total distribution of the duration is computed in (17).

$$T = \underbrace{\frac{4}{10}T}_{T_{acc}} + \frac{2}{10}T + \underbrace{\frac{4}{10}T}_{T_{dec}}, \tag{17}$$

where

$$T_{acc} = T_{dec} = 0.72 \text{ s.}$$

The acceleration time was compounded by three phases of the jerk time, as presents (18). When $\frac{T_{acc}}{2}$ the jerk value turns to zero, it corresponds to the time interval $T_{s1} - T_{s2}$ from Figure 1d.

$$T_{acc} = \underbrace{\frac{T_{acc}}{4}}_{T_{jerk}} + \frac{T_{acc}}{2} + \underbrace{\frac{T_{acc}}{4}}_{T_{-jerk}}, \tag{18}$$

where

$$T_{jerk} = 0.18 \text{ s.}$$

The derivative respect the time of speed is calculated using (3). The parameter α_d depends on θ_d, T_{jerk}, T, and the proportional constants of time γ and φ. The magnitude of the acceleration is obtained from (12), so, using (19), α_d is computed.

$$\alpha_d = \frac{\theta_d}{\gamma(1-\gamma)(1-\varphi)T^2} = \frac{12\pi \text{ } rad}{0.5832 \text{ } s^2} = 64.6418 \text{ rad/s}^2. \tag{19}$$

To estimate the value of the jerk, Equation (13) was used. Since the acceleration α_d has been obtained and the jerk interval T_{jerk} is known, it was possible to obtain the rate of change in acceleration in (20).

$$J_d = \frac{\alpha_d}{T_{jerk}} = \frac{64.6418 \text{ } rad/s^2}{0.18 \text{ } s} = 359.1212 \text{ rad/s}^3. \tag{20}$$

In Table 1 is presented all the parameters needed to generate the algorithms in order to compute the trajectory proposed.

Table 1. S-curve parameters used for implementation.

Parameters		Values
Desired position	θ_d	12π
Time of displacement	T	1.8 s
Time factor for acceleration time	γ	0.4
Time factor for jerk phase	φ	0.25
Acceleration time	T_{acc}	0.72 s
Deceleration time	T_{dec}	0.18 s
Velocity	ω_d	34.9065 rad/s
Acceleration	α_d	64.6418 rad/s^2
Jerk	J_d	359.1212 rad/s^3

4. Simulation and Results

The algorithm proposed in this paper has been compared with the results presented in [33], where a fourth order polynomial S-curve is developed, the input data has been displayed in Table 2. For simulation, two lengths of acceleration phase were used. By one hand, a wide stage of acceleration was chosen with a factor $\gamma = \frac{2}{5}$; on the other hand, a short stage of acceleration with a factor $\gamma = \frac{1}{5}$ was used to compare the jerk response of both motion profiles. Besides, two target positions were chosen and taken from simulation section of [33], $\theta_{d1} = \frac{2\pi}{3}$ and $\theta_{d2} = \frac{\pi}{3}$.

Table 2. Simulation results compared from the [33] method.

		$\gamma = \frac{2}{5}$		$\gamma = \frac{1}{5}$		Yi Fang and Wenhai Liu [33]	
Position (rad)	Initial point	0	0	0	0	0	0
	Final point	$\frac{2\pi}{3}$	$\frac{\pi}{6}$	$\frac{2\pi}{3}$	$\frac{\pi}{6}$	$\frac{2\pi}{3}$	$\frac{\pi}{6}$
Kinematics constraints	Velocity (rad/s)	2.319	0.5799	1.736	0.434	8	5
	Acceleration (rad/s^2)	5.137	1.2834	8.633	2.158	10	8
	Jerk (rad/s^3)	22.3	8.533	29.36	21.53	30	20

In Figure 6b, one can see the behavior of the velocity. For factor $\gamma = \frac{2}{5}$, the S-curve profile maintain symmetric intervals for acceleration-deceleration phases and maximum velocity. For $\gamma = \frac{1}{5}$ the acceleration-deceleration stages are shorter than the velocity phase. The position in Figure 6a adopts a different shape because of the acceleration-deceleration phase. Evaluating the response of the velocity profile for both factors, using an execution time of 1.5 s and a desired position of $\frac{2\pi}{3}$, the maximum speed for $\gamma = \frac{2}{5}$ is $\omega_{d1} = 2.319$ rad/s. On the other hand, the maximum velocity reached for the motion profile with a factor $\gamma = \frac{1}{5}$ is $\omega_{d2} = 1.736$ rad/s. The magnitude of the velocity obtained in [33] is bigger compared with the results obtained with the two factors proposed. The same occurred when the desired position was changed.

As mentioned before, factors $\gamma = \frac{1}{5}$ and $\gamma = \frac{2}{5}$ affect directly to the acceleration-deceleration phases. It means, when the length of the acceleration stage was short, the magnitude of the acceleration increased considerably. Figure 7a shows the behavior of the acceleration respect with each factor. Notice that, whether the acceleration-deceleration phases are short, the jerk tends to increase in magnitude as is displayed in Figure 7b. The jerk value was computed with respect to θ_d and T. The maximum jerk values were obtained for short acceleration-deceleration using $\gamma = \frac{1}{5}$ and $\theta_d = \frac{2\pi}{3}$. Unlike the jerk values obtained with the Yi Fang and Wenhai Liu algorithm $J_{max} = 30$ rad/s^3, the proposed algorithm performed a jerk value $J_{max} = 29.36$ rad/s^3. On the other hand, the acceleration stage in [33] is compounded by seven intervals, while the proposed method in this paper has three phases showing better simulation results. The length of the acceleration phase allows to reduce or increase the magnitude of the jerk, for a factor $\gamma < \frac{2}{5}$ the magnitude increases, while for a factor $\gamma = \frac{2}{5}$ the magnitude decreases, these factors can be chosen by the path planning designer.

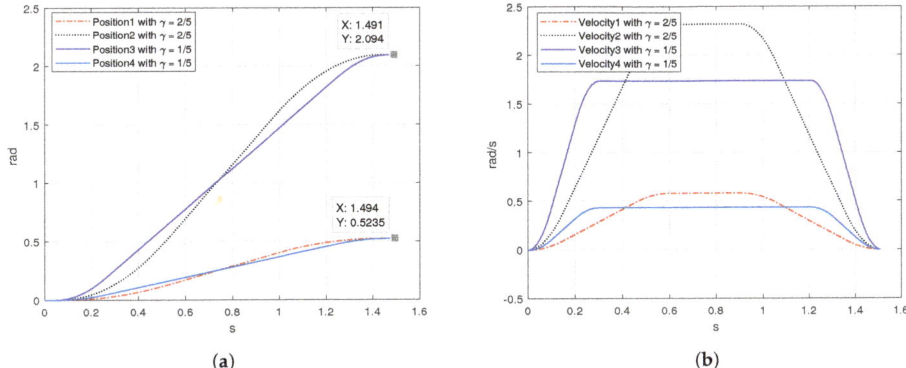

Figure 6. Position θ_d and S-curve velocity profile simulation with $\gamma = \frac{1}{5}$ and $\gamma = \frac{2}{5}$ factors (**a**) position, (**b**) velocity.

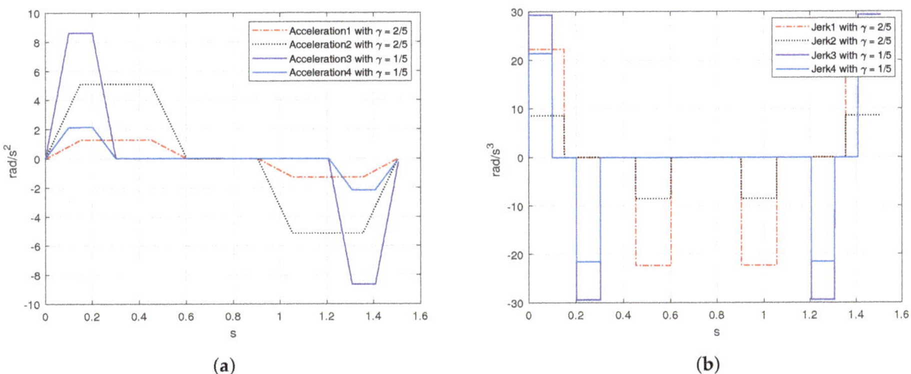

Figure 7. Acceleration and jerk simulation with $\gamma = \frac{1}{5}$ and $\gamma = \frac{2}{5}$ factors (**a**) acceleration, (**b**) jerk.

In this section, trapezoidal and S-curve velocity profiles were implemented to compare the response in real-time of the behavior of velocity and to measure the error in position of both motion profiles. A cylindrical load of 0.300 kg with inertia of 0.00011344 kg·m^2 was coupled to the shaft of a DC motor, as shows Figure 8, to obtain the experimental results. The motor has to compensate its movement even with the load to achieve the desired position θ_d following the trajectory computed by the motion profile.

4.1. Trapezoidal Velocity Profile

The trapezoidal velocity profile is the most used in industrial applications due to the ease of implementation since it consists of two linear equations describing the acceleration-deceleration phases, and a constant velocity stage [34]. The change in velocity was radical, so, the change in acceleration tended to infinity, mathematically speaking, but in real-world applications, the jerk permits an increment in residual vibrations, and damage in motors in certain period.

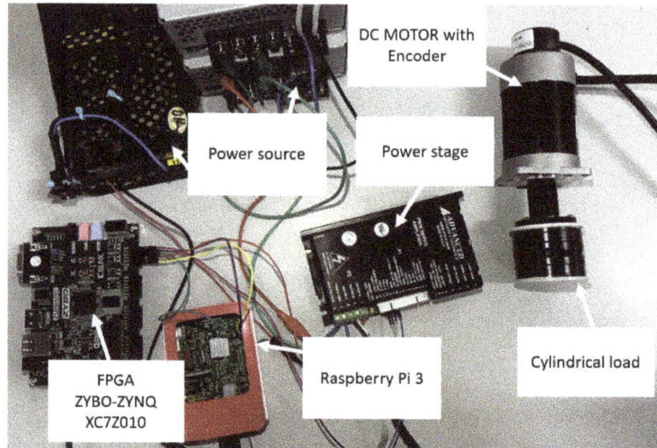

Figure 8. Low-cost platform based on FPGA-SoC.

The trapezoidal velocity profile presented in this section must reach a desired position $\theta_d = 12\pi$ rad in a total period of $T = 1.8$ s with a maximum velocity $\omega_{d_T} = 32$ rad/s. The trapezoidal velocity profile experimentally obtained is presented in Figure 9b.

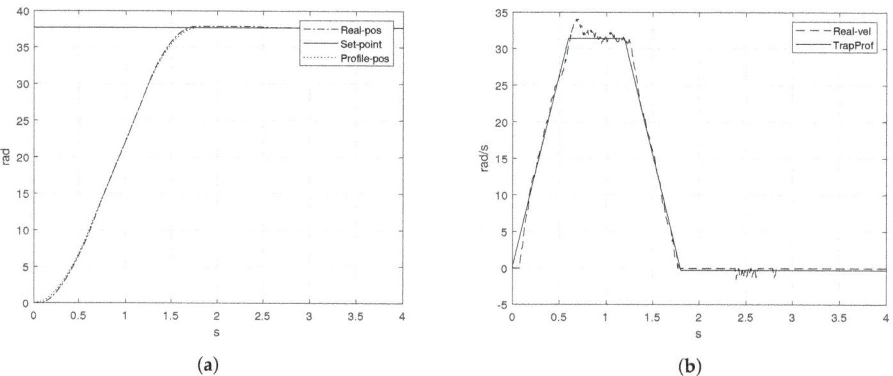

Figure 9. Trapezoidal velocity profile implementation (**a**) position, (**b**) velocity.

As the Figure 9b shows, the velocity follows the shape of the desired trapezoidal velocity profile with a disturbance when the speed has to be constant. The disturbance increased since the speed changed all of the sudden, so that the motor can not react instantly to follow the rate of change in position. The maximum peak of the real velocity goes to 34.3 rad/s although the desired position was reached, Figure 9a, there exists an error of $e_T = 0.18$ rad when the position should have reached the set-point in T. The error signal of the trapezoidal velocity is shown in Figure 10a. On the other hand, the maximum voltage required to achieve the maximum speed is about $u(t) = 1.39$ V and the control signal is presented in Figure 10b.

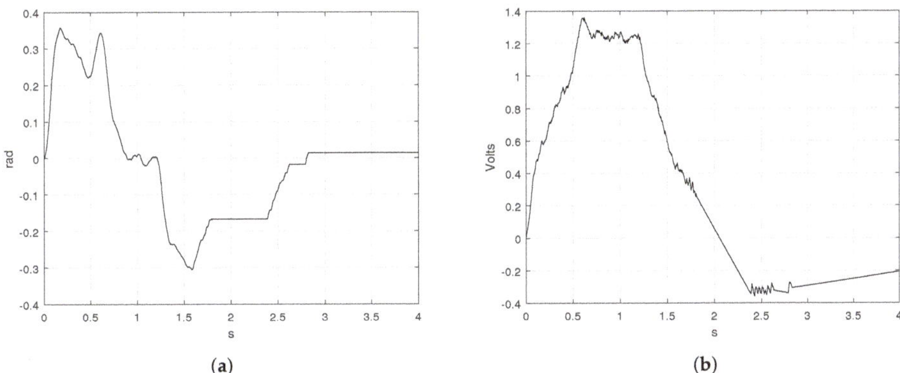

(a)   (b)

Figure 10. Error and control signals obtained from the trapezoidal velocity profile implementation (**a**) error position and (**b**) control signal.

4.2. S-Curve Velocity Profile Implementation

The implementation of the S-curve velocity profile was applied in the low-cost platform using the values presented in Table 1 and the methodology proposed in Section 3. Notice that the parameters proposed for the S-curve profile were similar to the trapezoidal velocity profile. The S-curve velocity profile obtained experimentally is presented in Figure 11b.

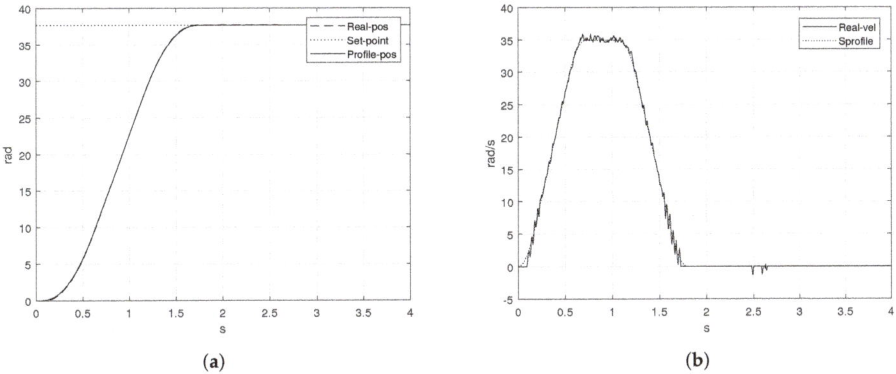

(a)   (b)

Figure 11. S-curve velocity profile implementation (**a**) position and (**b**) velocity.

The speed measured by the rotary encoder follows properly the speed computed by the algorithm. When the velocity reached the constant phase, the seven-segment velocity profile presented a smooth change in the velocity. Besides, the velocity showed in Figure 9b was lower in magnitude than the obtained by the S-curve profile. The value of θ_d was reached in the proposed time T. Figure 11a shows the behavior of the real position of the shaft of the motor, the real position is pretty similar to the computed position.

The error signal presented in Figure 12a exhibited an error when the position must be reached the set-point of $e_s = 0.04$ rad. The control signal is presented in Figure 12b, where the maximum voltage required for all the displacement is $u(t) = 1.5$ V.

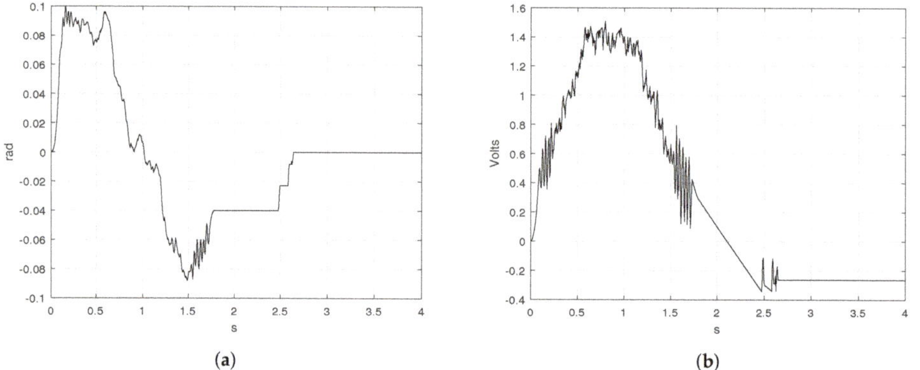

(a)

(b)

Figure 12. Error position and control signals obtained from the S-curve velocity profile implementation (**a**) error and (**b**) control signal.

The acceleration behavior is shown in Figure 13a. The acceleration obtained from the trapezoidal velocity profile shows an aggressive shape due to the acceleration profile is ideally an square signal. Experimentally, the acceleration value was $\alpha_T = \pm 52.3$ rad/s^2. The shape of the velocity profile was important to compensate the behavior of the acceleration and avoid discontinuities because of the jerk.

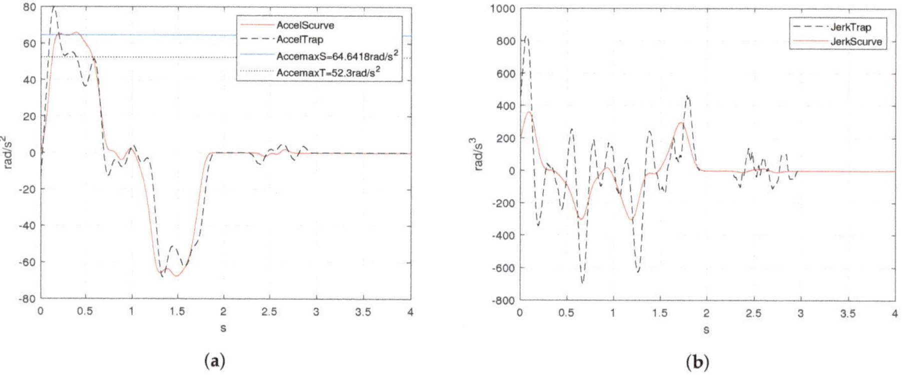

(a)

(b)

Figure 13. Acceleration and jerk of trapezoidal and S-curve motion profiles (**a**) acceleration and (**b**) jerk.

The S-curve velocity profile displays an smooth response in speed and acceleration. The acceleration value computed in (19) is $\alpha_d = 64.6418$ rad/s^2. The form of the acceleration computed by the S-curve velocity profile is smoother than the obtained by the trapezoidal. Besides, it maintains the acceleration value computed mathematically.

The mathematical value for the jerk derived from the trapezoidal velocity profile tended to infinity, but for experimental results, the rate of change in acceleration produced an increase in vibrations and the magnitude is relatively high for the acceleration and deceleration phases. For the S-curve velocity profile, the jerk profile is bounded by the designer in (20), so the computed jerk for the experiment was $J_d = 359.1212$ rad/s^3. Figure 13b presents the jerk behavior for both velocity profiles. One can prove that the response presented for the S-curve velocity profile was lower in magnitude than the trapezoidal velocity profile. This happens because of the third-degree polynomial proposed for the position profile.

Two positions desired were proposed to prove the negative velocity and displacement constraints in implementation. The values used in the algorithm are presented in Table 3.

Table 3. S-curve and inverse S-curve parameters used for implementation.

Parameters		Values
First Displacement		
Desired position	θ_{d_1}	12π rad
Time of displacement	T_1	1.8 s
Velocity	ω_{d_1}	34.9065 rad/s
Acceleration	α_{d_1}	64.6418 rad/s^2
Jerk	J_{d_1}	359.1212 rad/s^3
Second Displacement		
Desired position	θ_{d_2}	0 rad
Time of displacement	T_2	1.8 s
Velocity	ω_{d_2}	-34.9065 rad/s
Acceleration	α_{d_2}	-64.6418 rad/s^2
Jerk	J_{d_2}	-359.1212 rad/s^3

When the actuator has reached the first set-point, $\theta_{d_1} = 12\pi$, the shaft of the motor was maintained in that point until the new desired position was added. So that, $\theta_0 = \theta_{d_1}$, it means θ_0 is the last point measured by the encoder after $T_1 = 1.8$ s and was the initial position for the next displacement, T_1 is the total duration of the first movement. The new position to reach was $\theta_{d_2} = 0$ rad, the aim was to set the shaft of the motor till the origin using the same period of time $T_2 = 1.8$ s. The magnitude of the velocity was the same for both trajectories but the direction was different, so $\omega_{d_2} = -34.9065$ rad/s. The S-curve and the inverse S-curve is presented in Figure 14b. The position profile for both displacements has the shape of the S-curve profile, the position of Figure 14a is displayed by two different movements. The set-points are reached properly in the proposed time. The position measured by the encoder follows the computed position by the algorithm.

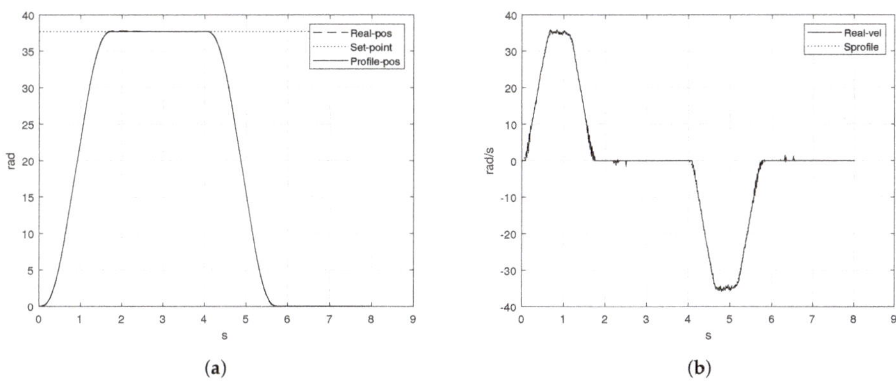

Figure 14. S-curve and inverse S-curve velocity profile implementation, (a) position, and (b) velocity.

The error signal is displayed in Figure 15, the first movement has a duration of $T_1 = 1.8$ s. When $t = T_1$, the error was about $e_s = 0.05$ rad, it means the position is near θ_{d_1}, after 0.47 s the actual error has been decreased to $e_s = 0.009$ rad.

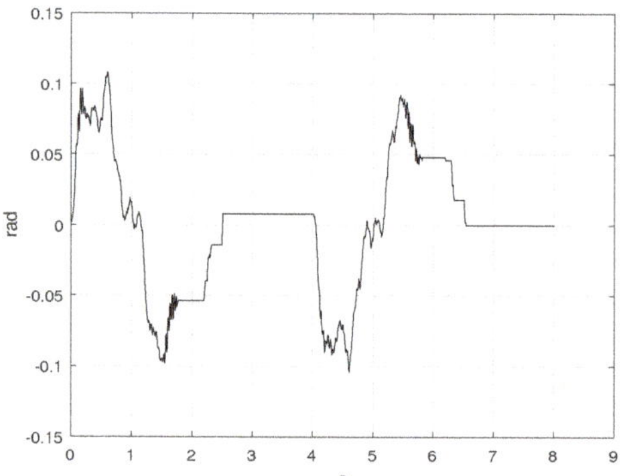

Figure 15. Error position signal of the S-curve and inverse S-curve implementation.

The new displacement started in $t = 4$ s. The length of the second movement was $T_2 = 1.8$ s. When $t = 5.8$ s, the present error had a magnitude of $e_{is} = 0.049$ rad, the error turned to zero ($e = 0$ rad) in $t = 6.5$ s according to the Figure 15.

5. Conclusions

In this paper, an S-curve velocity profile has been presented for motion control applications implemented in a low-cost platform compounded by a Raspberry Pi 3 and an FPGA. The velocity profile algorithm can be modified to generate a specific trajectory varying the parameters of implementation, such as the length of the acceleration and deceleration phases. Besides, given a desired position θ_d and the total duration of the movement T, one can obtain the magnitude of the speed ω_d, acceleration α_d and the jerk J_d to generate the S-curve velocity profile. The error obtained with the trapezoidal velocity profile was $e_T = 0.18$ rad when the duration of the movement is $T = 1.8$ s. The error obtained with the application of the S-curve velocity profile is about $e_s = 0.05$ rad in the same time.

Notice that the magnitude of the speed used in the S-curve was $\omega_d = 34.9065$ rad/s and $\omega_{d_T} = 32$ rad/s for the trapezoidal velocity profile. Despite the maximum speed reached for the trapezoidal profile is less in magnitude than the S-curve profile, the velocity behavior does not follow properly the computed speed by the algorithm because of the rate of change in acceleration. The S-curve velocity profile presents a smoother change in the position than the trapezoidal profile due to the third-degree polynomial proposed in the acceleration-deceleration stages.

Author Contributions: Conceptualization, J.R.G.-M. and J.R.-R.; methodology, J.R.G.-M.; software, J.R.G.-M.; validation, J.R.G.-M., E.E.C.-M. and J.R.-R.; formal analysis, J.R.G.-M.; investigation and visualization, J.R.G.-M., E.E.C.-M. and J.R.-R.; data curation, J.R.G.-M., E.E.C.-M. and J.R.-R.; writing—original draft preparation; writing—original draft, review, and editing, all the authors.

Funding: This research was funded by the "Consejo Nacional de Ciencia y Tecnología (CONACYT)" under the scholarship 778619.

Acknowledgments: We would like to thank the Graduate Studies Division from the Faculty of Engineering at Universidad Autónoma de Querétaro by allowing me to make Ph.D. studies.

Conflicts of Interest: The authors declare no conflict of interest.

References

1. Erkorkmaz, K.; Altintas, Y. High speed CNC system design. Part I: Jerk limited trajectory generation and quintic spline interpolation. *Int. J. Mach. Tools Manuf.* **2001**, *41*, 1323–1345. [CrossRef]
2. Martínez, J.R.G.; Reséndiz, J.R.; Prado, M.Á.M.; Miguel, E.E.C. Assessment of jerk performance s-curve and trapezoidal velocity profiles. In Proceedings of the 2017 XIII International Engineering Congress (CONIIN), Santiago de Queretaro, Mexico, 15–19 May 2017; pp. 1–7.

3.   Osornio-Rios, R.A.; de Jesús Romero-Troncoso, R.; Herrera-Ruiz, G.; Castañeda-Miranda, R.   FPGA implementation of higher degree polynomial acceleration profiles for peak jerk reduction in servomotors. *Robot. Comput.-Integr. Manuf.* **2009**, *25*, 379–392. [CrossRef]

4.   Rew, K.H.; Kim, K.S.   A closed-form solution to asymmetric motion profile allowing acceleration manipulation. *IEEE Trans. Ind. Electron.* **2010**, *57*, 2499–2506. [CrossRef]

5.   Kim Doang Nguyen, T.C.N. On Algorithms for Plannin S-curve Motion Profiles. *INTECH* **2008**, *5*, 99–106.

6.   Piazzi, A.; Visioli, A.   Global minimum-jerk trajectory planning of robot manipulators.   *IEEE Trans. Ind. Electron.* **2000**, *47*, 140–149. [CrossRef]

7.   Zhang, Y.; He, L.; Luo, J.; Tan, H.   Complete framework of jerk-level inverse-free solutions to inverse kinematics of redundant robot manipulators. In Proceedings of the 2016 35th Chinese Control Conference, Chengdu, China, 27–29 July 2016; pp. 4717–4722. [CrossRef]

8.   Jeon, J.K. An efficient acceleration for fast motion of industrial robots. In Proceedings of the IECON'95-21st Annual Conference on IEEE Industrial Electronics, Orlando, FL, USA, 6–10 November 1995; Volume 2, pp. 1336–1341. [CrossRef]

9.   Lloyd, J.; Vincent, H. Trajectory Generation for Sensor-Driven and Time-Varying Tasks. *Int. J. Robot. Res.* **1993**, *12*, 380–393. [CrossRef]

10.   Lewin, C. *Mathematics of Motion Control Profiles*; Performance Motion Devices, Inc.: Westford, MA, USA, 2007; pp. 1–5.

11.   Heo, H.J.; Son, Y.; Kim, J.M. A Trapezoidal Velocity Profile Generator for Position Control Using a Feedback Strategy. *Energies* **2019**, *12*, 1222. [CrossRef]

12.   Macfarlane, S.; Croft, E. Jerk-bounded manipulator trajectory planning: Design for real-time applications. *IEEE Trans. Robot. Autom.* **2003**, *19*, 42–52. [CrossRef]

13.   Siciliano, B.; Sciavicco, L.; Villani, L.; Oriolo, G. *Robotics: Modelling, Planning and Control*; Springer Science & Business Media: Berlin/Heidelberg, Germany, 2010.

14.   Bai, Y.; Chen, X.; Sun, H.; Yang, Z. Time-Optimal Freeform S-curve Profile under Positioning Error and Robustness Constraints. *IEEE/ASME Trans. Mechatron.* **2018**, *4435*, 1–11. [CrossRef]

15.   Jeon, J. A Generalized Approach for the Acceleration and Deceleration of Industrial Robots and CNC Machine Tools. *IEEE Trans. Ind. Electron.* **2000**, *47*, 133–139. [CrossRef]

16.   Martínez-Prado, M.A.; Rodríguez-Reséndiz, J.; Gómez-Loenzo, R.A.; Herrera-Ruiz, G.; Franco-Gasca, L.A. An FPGA-Based Open Architecture Industrial Robot Controller. *IEEE Access* **2018**, *6*, 13407–13417. [CrossRef]

17.   Ni, H.; Zhang, C.; Ji, S.; Hu, T.; Chen, Q.; Liu, Y.; Wang, G. A Bidirectional Adaptive Feedrate Scheduling Method of NURBS Interpolation Based on S-Shaped ACC/DEC Algorithm. *IEEE Access* **2018**, *6*, 63794–63812. [CrossRef]

18.   Ogata, K.; Yang, Y. *Modern Control Engineering*; Prentice Hall: Delhi, India, 2002; Volume 4.

19.   Nise, N.S. *Control Systems Engineering, (With CD)*; John Wiley & Sons: Hoboken, NJ, USA, 2007.

20.   Sabanovic, A.; Ohnishi, K. *Motion Control Systems*; John Wiley & Sons: Hoboken, NJ, USA, 2011.

21.   Ding, H.; Wu, J. Point-to-point motion control for a high-acceleration positioning table via cascaded learning schemes. *IEEE Trans. Ind. Electron.* **2007**, *54*, 2735–2744. [CrossRef]

22.   Jeong, S.Y.; Choi, Y.J.; Park, P.; Choi, S.G. Jerk limited velocity profile generation for high speed industrial robot trajectories. *IFAC Proc. Vol.* **2005**, *38*, 595–600. [CrossRef]

23.   Wang, B.; Liu, Q.; Zhou, L.; Zhang, Y.; Li, X.; Zhang, J. Velocity profile algorithm realization on FPGA for stepper motor controller. In Proceedings of the 2011 2nd International Conference on Artificial Intelligence, Management Science and Electronic Commerce, Zhengzhou, China, 8–10 August 2011; pp. 6072–6075. [CrossRef]

24.   Kei, T.W.; Mang, V.; Un, C.S. Design of s-curve direct landing position control system for elevator using microcontroller. In Proceedings of the World Congress on Engineering and Computer Science, San Francisco, CA, USA, 24–26 October 2012; Volume II, pp. 24–27.

25.   Angeles, J.; López-Cajún, C.S.   *Optimization of Cam Mechanisms*; Springer Science & Business Media: Berlin/Heidelberg, Germany, 2012; Volume 9.

26.   Ha, C.W.; Lee, D. Analysis of Embedded Prefilters in Motion Profiles. *IEEE Trans. Ind. Electron.* **2018**, *65*, 1481–1489. [CrossRef]

27.   Li, H.; Le, M.; Gong, Z.; Lin, W. Motion profile design to reduce residual vibration of high-speed positioning stages. *IEEE/ASME Trans. Mechatron.* **2009**, *14*, 264–269.

28. Gurocak, H. *Industrial Motion Control: Motor Selection, Drives, Controller Tuning, Applications*; John Wiley & Sons: Hoboken, NJ, USA, 2015.

29. Padilla-Garcia, E.A.; Rodriguez-Angeles, A.; Reséndiz, J.R.; Cruz-Villar, C.A. Concurrent Optimization for Selection and Control of AC Servomotors on the Powertrain of Industrial Robots. *IEEE Access* **2018**, *6*, 27923–27938. [CrossRef]

30. Xueshan, Y.; Xiaozhai, Q.; Lee, G.C.; Tong, M.; Jinming, C. Jerk and jerk sensor. In Proceedings of the 14th World Conference on Earthquake Engineering, Beijing, China, 12–17 October 2008.

31. Bearee, R.; Barre, P.J.; Hautier, J.P. Vibration reduction abilities of some jerkcontrolled movement laws for industrial machines. In Proceedings of the 16th IFAC World Congress, Prague, Czech Republic, 3–8 July 2005.

32. *Analog Servo Drive*; Rev. 2.0.1; Advanced Motion Controls: Camarillo, CA, USA, 2011.

33. Fang, Y.; Hu, J.; Liu, W.; Shao, Q.; Qi, J.; Peng, Y. Smooth and time-optimal S-curve trajectory planning for automated robots and machines. *Mech. Mach. Theory* **2019**, *137*, 127–153. [CrossRef]

34. Yoon, H.; Chung, S.; Kang, H.; Hwang, M. Trapezoidal Motion Profile to Suppress Residual Vibration of Flexible Object Moved by Robot. *Electronics* **2019**, *8*, 30. [CrossRef]

MDPI

St. Alban-Anlage 66

4052 Basel

Switzerland

Tel. +41 61 683 77 34

Fax +41 61 302 89 18

www.mdpi.com

Electronics Editorial Office

E-mail: electronics@mdpi.com

www.mdpi.com/journal/electronics